Glencoe

Geometry

Integration
Applications
Connections

Practice Workbook

McGraw-Hill

New York, New York Columbus, Ohio Woodland Hills, California Peoria, Illinois

Glencoe/McGraw-Hill

A Division of The ***McGraw·Hill*** *Companies*

Send all inquiries to:
Glencoe/McGraw-Hill
8787 Orion Place
Columbus, OH 43240-4027

Geometry
Practice Workbook

ISBN: 0-02-825322-1

9 10 066 04 03 02 01 00

Contents

To the Student

This *Practice Workbook* gives you additional practice for the concept exercises in each lesson. The practice exercises are designed to aid your study of mathematics by reinforcing important mathematical skills needed to succeed in the everyday world. The material is organized by chapter and lesson with one practice worksheet for every lesson in *Geometry.*

Always keep your *Practice Workbook* handy. Along with your textbook, daily homework, and class notes, the completed *Practice Workbook* can help you in reviewing for quizzes and tests.

To the Teacher

Answers to each practice worksheet are found in the *Practice Masters Booklet* and also in the Teacher's Wraparound Edition of *Geometry.*

NAME______________________ DATE ____________

Practice

Student Edition
Pages 6–11

Integration: Algebra
The Coordinate Plane

Graph each point on the coordinate plane at the right.

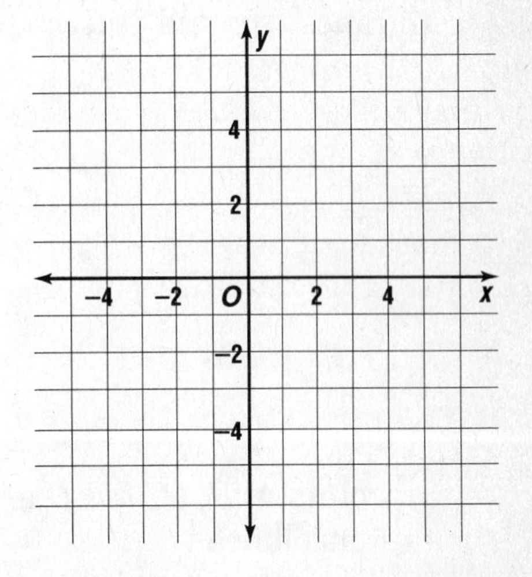

1. $A(3, -5)$ **2.** $B(5, 0)$

3. $C(-2, -1)$ **4.** $D(0, -6)$

5. $E(4, 3)$ **6.** $F(-4, 3)$

In the figure at right, all of the segments lie on the gridlines of a coordinate plane. Determine the ordered pair that represents each point.

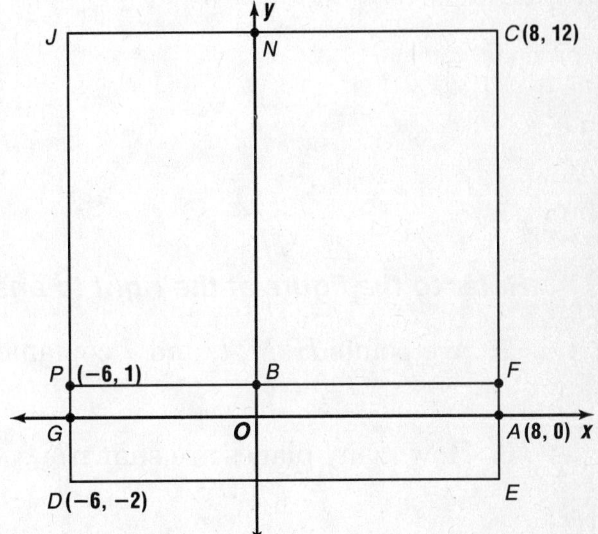

7. J **8.** G

9. B **10.** N

11. F **12.** E

Points G (1, −1) and H (3, 3) lie on the graph of y = 2x − 3. Determine whether the each point is collinear with G and H.

13. $I(0, -3)$ **14.** $J(2, 1)$

15. $K(-3, -8)$ **16.** $L(5, 7)$

17. $M(10, 16)$ **18.** $N(25, 48)$

19. $P(-10, -23)$ **20.** $Q(-35, -67)$

NAME ______________________ DATE ____________

Practice

Points, Lines, and Planes

Draw and label a figure for each relationship.

1. Lines ℓ, m and j intersect at P.

2. Plane $\mathcal{N}$ contains line ℓ.

3. Points A, B, C, and D are noncollinear.

4. Points A, B, C, and D are noncoplanar.

Refer to the figure at the right to answer each question.

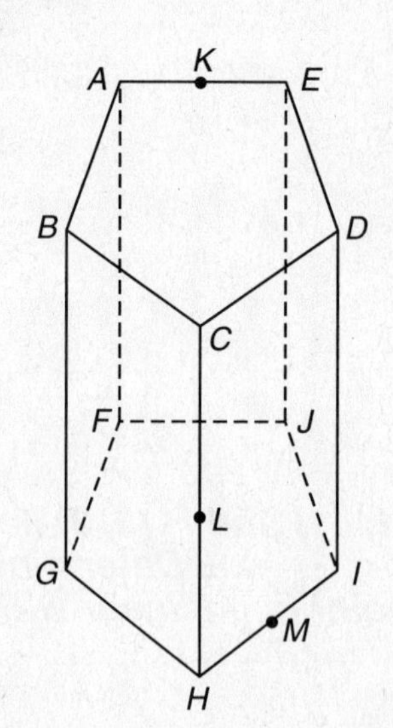

5. Are points H, M, I, and J coplanar?

6. How many planes are shown?

7. Name the intersection of planes ABG and CHG.

8. Name the intersection of plane ABC and $\overleftrightarrow{HL}$.

9. Which segments are contained in all three of the planes GFH, CDI, and EDI?

10. List the Possibilities Tim can choose from a tan shirt, a blue shirt, and a green shirt. He can choose from black slacks or blue jeans. He can choose from a windbreaker, a sweatshirt, or a jacket. How many different outfits can he wear if he will not wear the sweatshirt with the black slacks?

1–3

NAME_______________________________________ DATE ______________

Practice

Student Edition
Pages 19–25

Integration: Algebra
Using Formulas

Find the perimeter and area of each rectangle.

1.

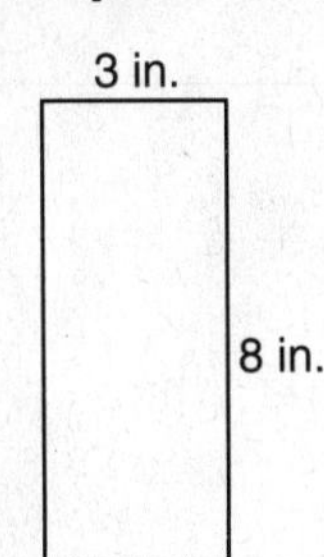

2.

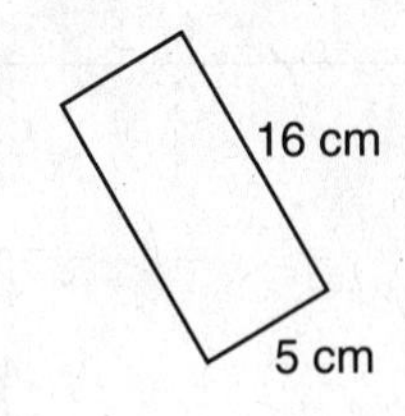

3.

18.3 mi

4.7 mi

Find the missing measure in each formula.

4. $\ell = 4, w = 2, P = \underline{?}$

5. $A = 32, w = 6.25, \ell = \underline{?}$

6. $P = 27, \ell = 8, w = \underline{?}$

7. $\ell = 3\frac{1}{2}, A = 5\frac{1}{4}, w = \underline{?}$

8. $w = 7, \ell = 12, P = \underline{?}$

9. $w = 15.5, P = 81, \ell = \underline{?}$

Find the maximum area for the given perimeter of a rectangle. State the length and width of the rectangle.

10. 25 mm

11. 18 yards

12. 42 cm

Practice

Measuring Segments

Given that B is between A and C, find each missing measure.

1. $AB = 5.3$, $BC = \underline{?}$, $AC = 6.7$

2. $AB = 21$, $BC = 4.3$, $AC = \underline{?}$

3. $AB = \underline{?}$, $BC = 18.9$, $AC = 23$

4. $AB = 6\frac{3}{4}$, $BC = \underline{?}$, $AC = 10$

If B is between A and C, find the value of x and the measure of BC.

5. $AB = 3x$, $BC = 5x$, $AC = 8$

6. $AB = 3(x + 7)$, $BC = 2(x - 3)$, $AC = 50$

Refer to the coordinate plane at the right to find each measure. Round your answers to the nearest hundredth.

7. AB

8. BD

9. AE

10. CE

11. AD

12. BE

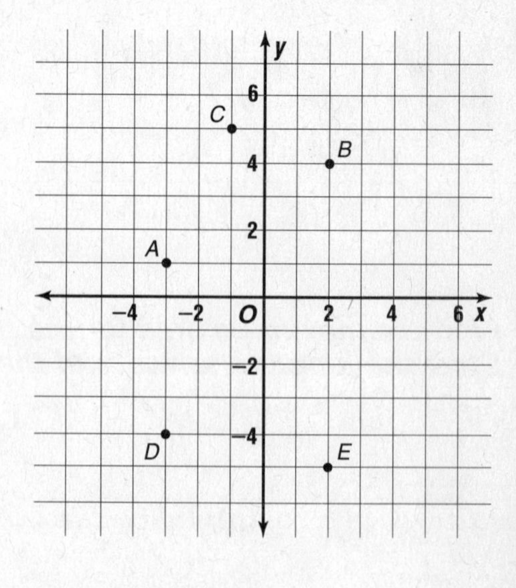

Midpoints and Segment Congruence

Refer to the figure below for Exercises 1–8 to determine whether each statement is true or false.

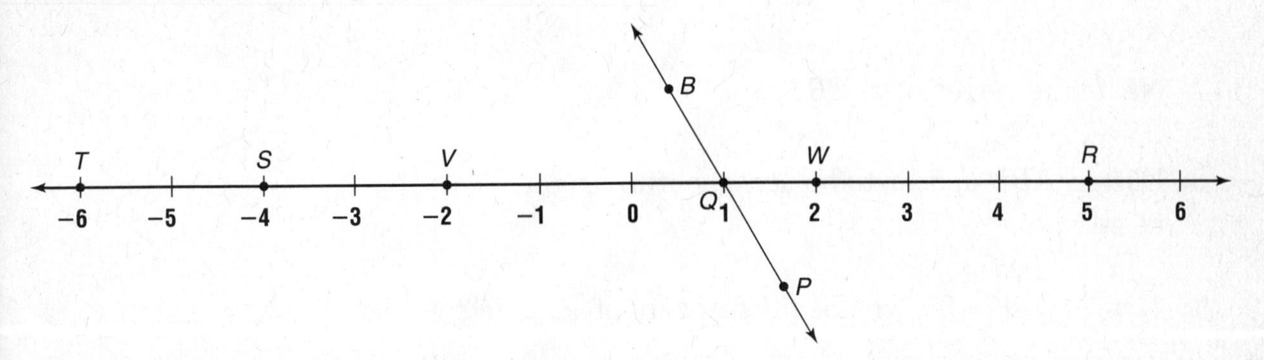

1. $\overline{PB}$ bisects $\overline{RS}$.

2. S is the midpoint of $\overline{TV}$.

3. $\overline{SV} \cong \overline{TS}$

4. V is the midpoint of $\overline{TW}$.

5. W bisects $\overline{VR}$.

6. $\overline{WR} \cong \overline{QV}$

7. $\overline{SW}$ is longer than $\overline{VR}$.

8. $VW \leq TV$

Given the coordinate of one endpoint of $\overline{AB}$ and its midpoint M, find the coordinates of the other endpoint.

9. $A(0, 9), M(2, 5)$

10. $B(-5, 1), M(1, -1)$

11. $A(-2, 3)\ M(0.5, 0.5)$

12. $A(4, 2), M(-2, 10)$

In the figure at the right, $\overline{WY}$ bisects $\overline{UV}$ at Y and $\overline{UY}$ bisects $\overline{TW}$ at X. For each situation, find the value of x and the measure of the indicated segment.

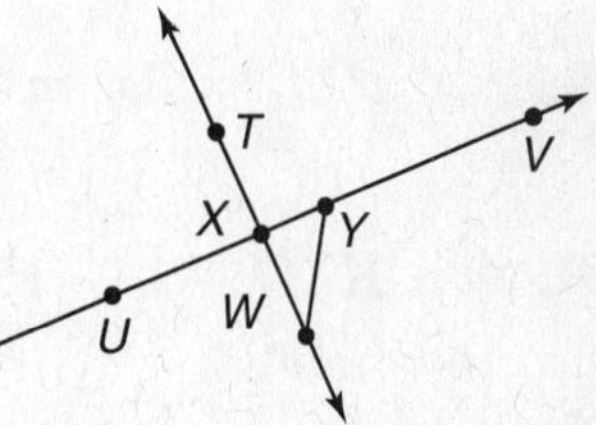

13. $UY = 4x - 3, YV = x; UV$

14. $UV = x + 6, UY = x - 1; YV$

15. $TX = 2x + 1, XW = x + 7; TW$

16. $WX = x + 5, TW = 4x + 5: TX$

Practice

Exploring Angles

Refer to the figure at the right to answer each question.

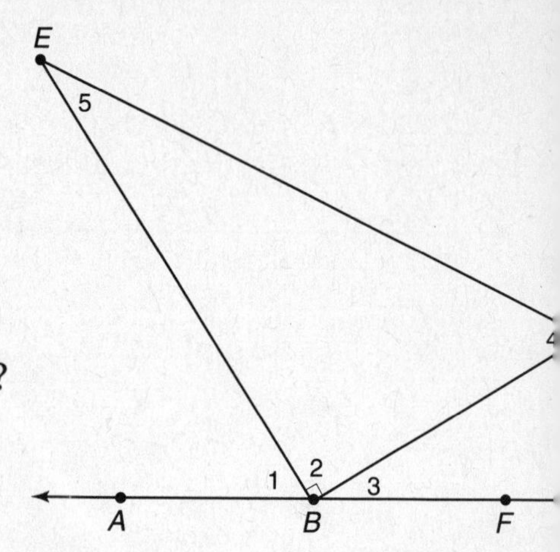

1. Give another name for $\angle 1$.

2. Name the vertex of $\angle EBD$.

3. Does $\angle ABE$ appear to be acute, obtuse, right, or straight?

4. If $m\angle DBF = 35$, what is the measure of $\angle EBF$?

5. Name a pair of opposite rays.

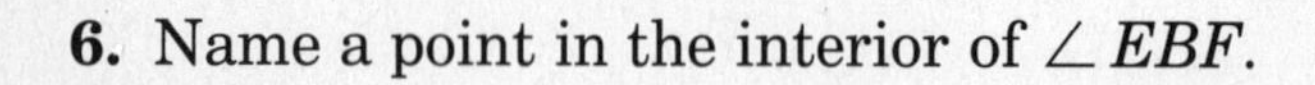

6. Name a point in the interior of $\angle EBF$.

7. Name three angles with $\overrightarrow{BE}$ as a side?

In the figure, $\overrightarrow{XP}$ and $\overrightarrow{XT}$ are opposite rays and $\overrightarrow{XQ}$ bisects $\angle PXS$. For each situation, find the value of x and the measure of the indicated angle.

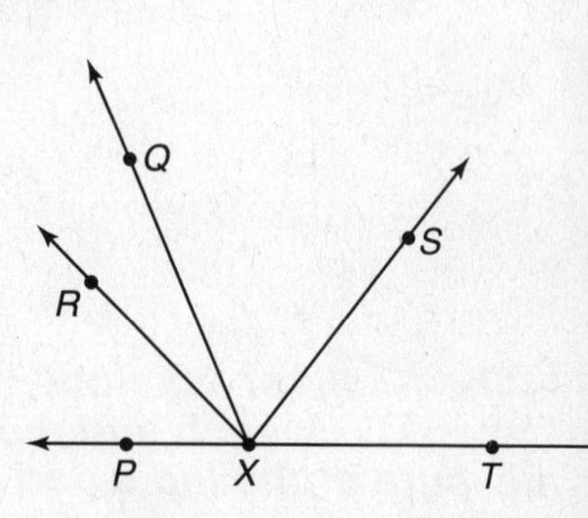

8. $m\angle SXT = 4x + 1$, $m\angle QXS = 2x - 2$, $m\angle QXT = 125$; $m\angle QXS$

9. $m\angle PXR = 3x$, $m\angle RXT = 5x + 20$, $m\angle RXT$

10. $m\angle RXQ = x + 15$, $m\angle RXS = 5x - 7$, $m\angle QXS = 3x + 5$; $m\angle RXS$

11. $m\angle RXQ = 2x + 7$, $m\angle RXP = 3x - 11$, $m\angle PXS = x + 37$; $m\angle QXS$

12. $m\angle TXS = x + 3$, $m\angle SXR = 2x + 9$, $m\angle RXP = 4x - 7$; $m\angle PXS$

NAME________________________ DATE ____________

1–7

Practice

Student Edition
Pages 53–60

Angle Relationships

Refer to the figure at the right to answer each question.

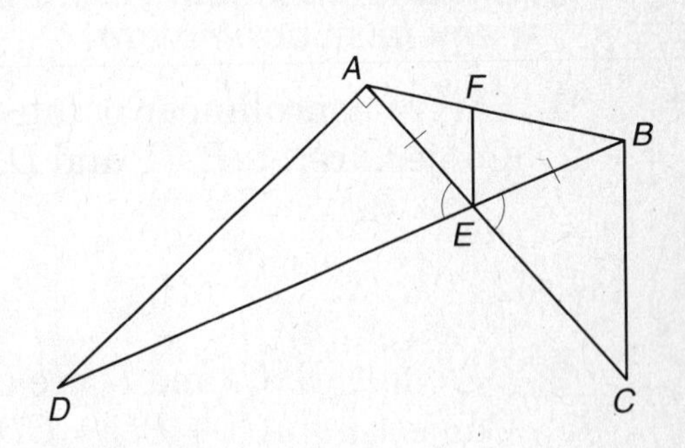

1. Name a pair of vertical angles.

2. Can you assume that $\angle EFB$ is a right angle from the figure?

3. Which angle is supplementary to $\angle FEB$?

4. Can you assume $\overline{AE} \cong \overline{BE}$?

5. Can you assume that F bisects $\overline{AB}$ from the figure?

6. Name an angle adjacent, but not supplementary, to $\angle DEA$.

Find the value of x and m∠ABC.

7.

A
B
(4x + 50)°
(2x + 60)°
C

8.

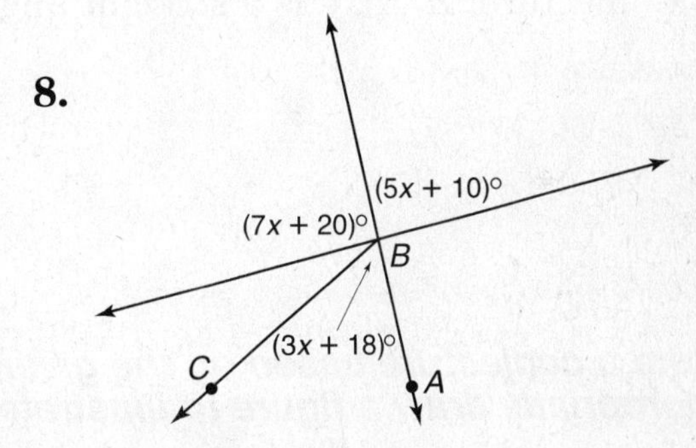

For each figure, find the value of x. Then determine if $\overline{AB} \perp \overline{CD}$.

9.

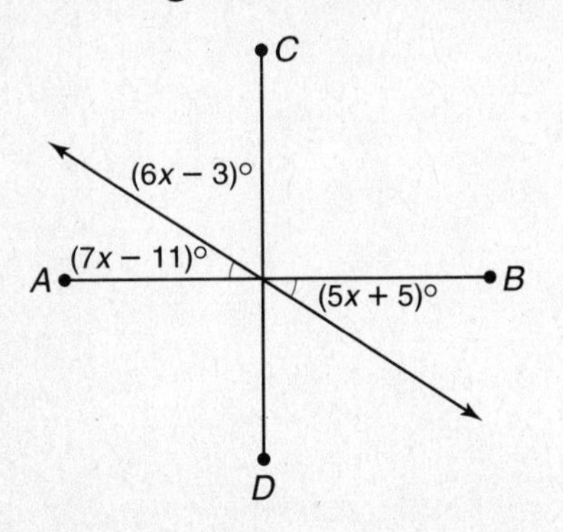

10.

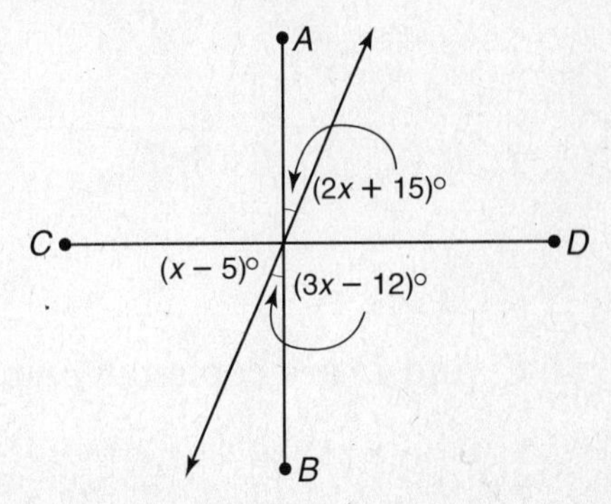

Practice

Inductive Reasoning and Conjecturing

Determine if the conjecture is true or false based on the given information. Explain your answer and give a counterexample for any false conjecture.

1. Given: noncollinear points A, B, C, and D
Conjecture: A, B, C, and D are coplanar.

2. Given: A, B, C, and D are collinear points.
Conjecture: $AB + BC + CD = AD$

3. Given: $\angle A$ and $\angle B$ are right angles.
Conjecture: $\angle A \cong \angle B$

4. Given: Point C between H and V.
Conjecture: $\angle HCV$ is a straight angle.

Write a conjecture based on the given information. If appropriate, draw a figure to illustrate your conjecture.

5. $\overline{AB}$, $\overline{CD}$, and $\overline{EF}$ intersect at X.

6. $\angle MNO$ and $\angle PNO$ are adjacent angles.

7. $A\ (3, 1)$, $B\ (3, -5)$, $C\ (3, 7)$.

8. A, B, C, and D are coplanar points.

2–2

NAME ______________________ DATE __________

Practice

Student Edition
Pages 76–83

If-Then Statements and Postulates

Identify the hypothesis and conclusion of each conditional statement.

1. If $3x - 1 = 7$, then $x = 2$.

2. If Carl scores 85%, then he passes.

Write each conditional statement in if-then form.

3. All students like vacations.

4. The game will be played provided it doesn't rain.

Write the converse of each conditional. Determine if the converse is true or false. It if is false, give a counterexample.

5. If it rains, then it is cloudy.

6. If x is an even number, then x is divisible by 2.

In the figure, P, Q, R, and S are in plane $\mathcal{N}$. Use the postulates you have learned to determine whether each statement is true or false.

7. R, S, and T are collinear.

8. There is only one plane that contains all the points R, S, and Q.

9. $\angle PQT$ lies in plane $\mathcal{N}$.

10. $\angle SPR$ lies in plane $\mathcal{N}$.

11. If X and Y are two points on line m, then $\overleftrightarrow{XY}$ intersects plane $\mathcal{N}$ at P.

12. Point K is on plane $\mathcal{N}$.

13. $\mathcal{N}$ contains $\overline{RS}$.

14. T lies in plane $\mathcal{N}$.

15. R, P, S, and T are coplanar.

16. ℓ and m intersect.

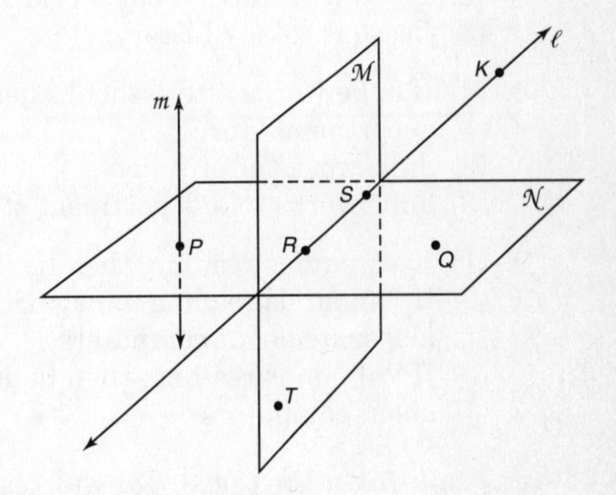

Deductive Reasoning

Determine if a valid conclusion can be reached from the two true statements using the Law of Detachment or the Law of Syllogism. If a valid conclusion is possible, state it and the law that is used. If a valid conclusion does not follow, write no conclusion.

1. If Jim is a Texan, then he is an American.
Jim is a Texan.

2. If Spot is a dog, then he has four legs.
Spot has four legs.

3. If Rachel lives in Tampa, then Rachel lives in Florida.
If Rachel lives in Florida, then Rachel lives in the United States.

4. If October 12 is a Monday, then October 13 is a Tuesday.
October 12 is a Monday.

5. If Henry studies his algebra, then he passes the test.
If Henry passes the test, then he will get a good grade.

Determine if statement (3) follows from statements (1) and (2) by the Law of Detachment or the Law of Syllogism. If it does, state which law was used. If it does not, write invalid.

6. (1) If the measure of an angle is greater than 90, then it is obtuse.
(2) $m\angle T$ is greater than 90.
(3) $\angle T$ is obtuse.

7. (1) If Pedro is taking history, then he will study about World War II.
(2) Pedro will study about World War II.
(3) Pedro is taking history.

8. (1) If Julie works after school, then she works in a department store.
(2) Julie works after school.
(3) Julie works in a department store.

9. (1) If William is reading, then he is reading a magazine.
(2) If William is reading a magazine, then he is reading a magazine about computers.
(3) If William is reading, then he is reading a magazine about computers.

10. Look for a Pattern Tanya likes to burn candles. She has found that, once a candle has burned, she can melt 3 candle stubs, add a new wick, and have one more candle to burn. How many total candles can she burn from a box of 15 candles?

NAME ______________________ DATE ____________

Practice

Integration: Algebra
Using Proof in Algebra

Name the property of equality that justifies each statement.

1. If $m\angle A = m\angle B$, then $m\angle B = m\angle A$.

2. If $x + 3 = 17$, then $x = 14$.

3. $xy = xy$

4. If $7x = 42$, then $x = 6$.

5. If $XY - YZ = XM$, then $XM + YZ = XY$.

6. $2(x + 4) = 2x + 8$.

7. If $m\angle A + m\angle B = 90$, and $m\angle A = 30$, then $30 + m\angle B = 90$.

8. If $x = y + 3$ and $y + 3 = 10$, then $x = 10$.

Complete each proof by naming the property that justifies each statement.

9. Prove that if $2(x - 3) = 8$, then $x = 7$.
Given: $2(x - 3) = 8$
Prove: $x = 7$
Proof:

Statements	Reasons
a. $2(x - 3) = 8$	**a.** ______________
b. $2x - 6 = 8$	**b.** ______________
c. $2x = 14$	**c.** ______________
d. $x = 7$	**d.** ______________

10. Prove that if $3x - 4 = \frac{1}{2}x + 6$, then $x = 4$.
Given: $3x - 4 = \frac{1}{2}x + 6$
Prove: $x = 4$
Proof:

Statements	Reasons
a. $3x - 4 = \frac{1}{2}x + 6$	**a.** ______________
b. $\frac{5}{2}x - 4 = 6$	**b.** ______________
c. $\frac{5}{2}x = 10$	**c.** ______________
d. $x = 4$	**d.** ______________

Practice

Verifying Segment Relationships

Complete each proof.

1. Given: $AD = 2AB + BC$
Prove: $\overline{AB} \cong \overline{CD}$
Proof:

Statements	**Reasons**
a. $AD = 2AB + BC$	**a.** ________
b. $AD = AB + BC + CD$	**b.** ________
c. $2AB + BC = AB + BC + CD$	**c.** ________
d. $AB = CD$	**d.** ________
e. $\overline{AB} \cong \overline{CD}$	**e.** ________

2. Given: B is between A and D.
C is between A and D.
Prove: $AB + BD = AC + CD$
Proof:

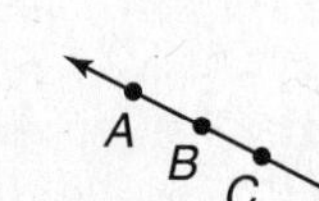

Statements	**Reasons**
a. B is between A and D. C is between A and D.	**a.** ________
b. $AB + BD = AD$	**b.** ________
c. $AC + CD = AD$	**c.** ________
d. $AD = AC + CD$	**d.** ________
e. $AB + BD = AC + CD$	**e.** ________

Write a two-column proof.

3. Given: B is the midpoint of $\overline{AC}$.
Prove: $AB + CD = BD$
Proof:

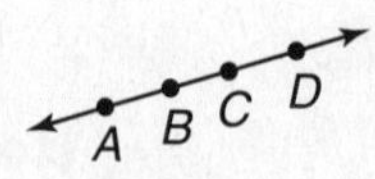

Statements	**Reasons**

NAME ______________________ DATE ____________

Practice

Student Edition
Pages 107–114

Verifying Angle Relationships

Complete each statement if m∠BGC = 43 and m∠DGE = 56.

1. $\angle FGA \cong$ _?_

2. $\angle BGF$ and _?_ are supplementary.

3. $m\angle CGD =$ _?_

4. $m\angle AGF =$ _?_

5. $\angle EGC$ and _?_ are supplementary.

6. $m\angle AGB =$ _?_

7. $m\angle AGC =$ _?_

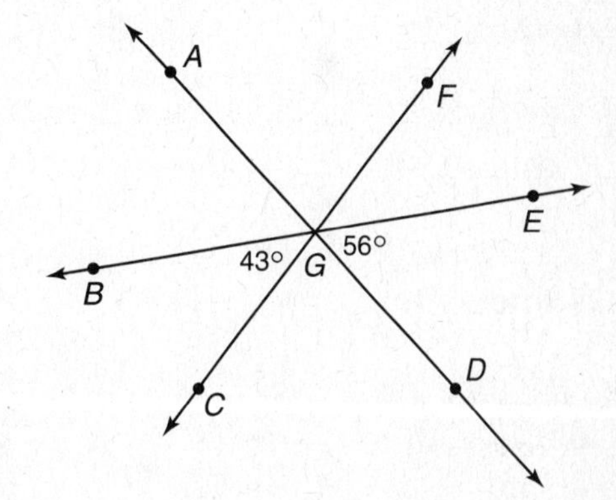

Write a two-column proof.

8. Given: $\angle AEC \cong \angle DEB$
Prove: $\angle AEB \cong \angle DEC$
Proof:

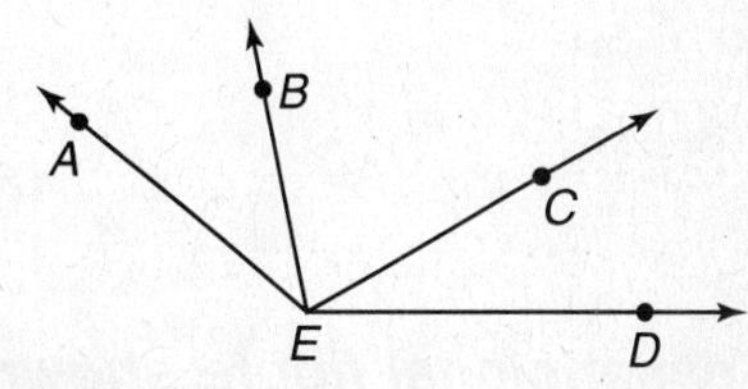

Statements	Reasons

Practice

Parallel Lines and Transversals

State the transversal that forms each pair of angles. Then identify the special name for the angle pair.

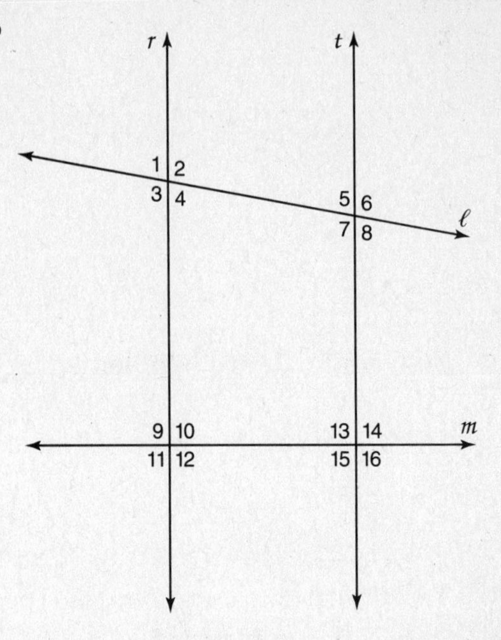

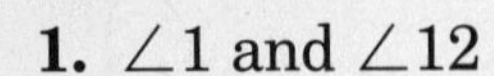

1. ∠1 and ∠12

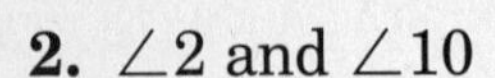

2. ∠2 and ∠10

3. ∠4 and ∠9

4. ∠6 and ∠3

5. ∠14 and ∠10

6. ∠7 and ∠13

The three-dimensional figure shown at the right is called a right pentagonal prism.

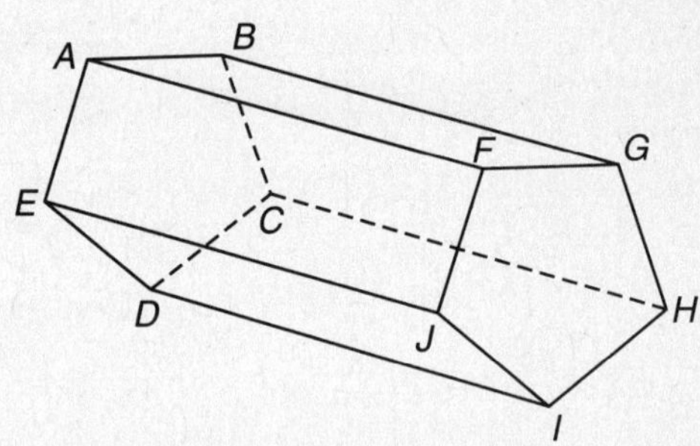

7. Identify all segments joining points marked in plane JIH that appear to be skew to $\overline{EB}$.

8. Which segments seem parallel to $\overline{BG}$?

9. Which segments seem parallel to $\overline{GH}$?

10. Identify all planes that appear parallel to plane FGH.

11. **Draw a Diagram** At a town's bicentennial celebration, men dressed up as settlers and tipped their hats whenever they met another man. At a town meeting, ten men were present. How many times were pairs of hats tipped as two men met for the first time?

NAME______________________ DATE __________

Practice

Angles and Parallel Lines

In the figure, $\ell \parallel m$. Find the measure of each angle.

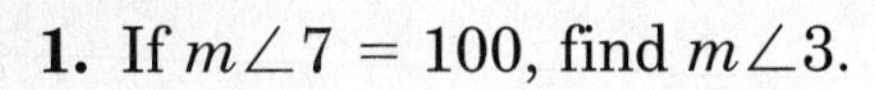

1. If $m\angle 7 = 100$, find $m\angle 3$.

2. If $m\angle 7 = 95$, find $m\angle 6$.

3. If $m\angle 1 = 120$, find $m\angle 5$.

4. If $m\angle 4 = 20$, find $m\angle 7$.

5. If $m\angle 3 = 140$, find $m\angle 8$.

6. If $m\angle 4 = 30$, find $m\angle 1$.

7. If $m\angle 4 = 40$, find $m\angle 2$.

8. If $m\angle 7 = 125$, find $m\angle 4$.

9. If $\ell \perp t$, find $m\angle 3$.

10. If $m\angle 1 + m\angle 3 = 230$, find $m\angle 6$.

In the figure, $s \parallel t$. Find the measure of each angle.

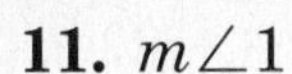

11. $m\angle 1$

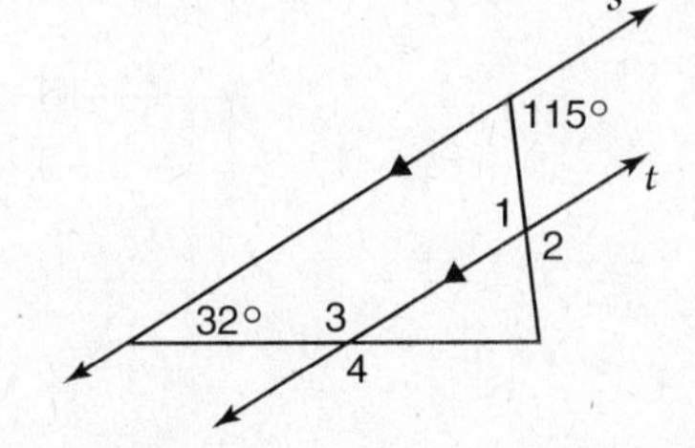

12. $m\angle 2$

13. $m\angle 3$

14. $m\angle 4$

15. In the figure, $r \parallel s$. Find the value of x.

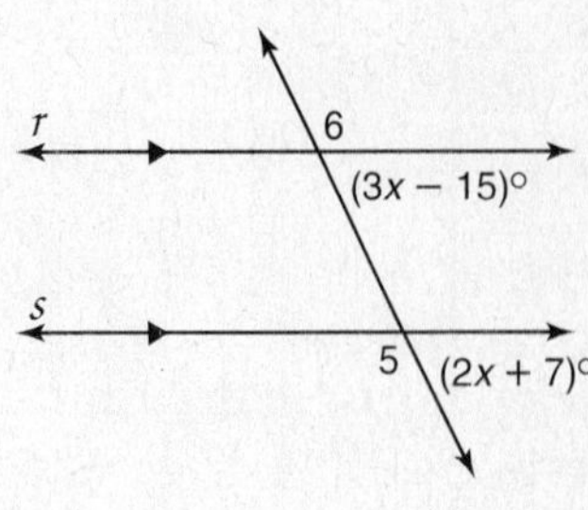

NAME ______________________________ DATE ____________

Practice

Integration: Algebra
Slopes of Lines

Determine the slope of each line named below.

1. a

2. b

3. c

4. d

5. any line parallel to b

6. any line perpendicular to d

Graph the line that satisfies each description.

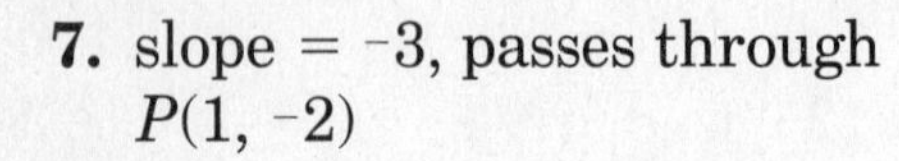

7. slope = -3, passes through $P(1, -2)$

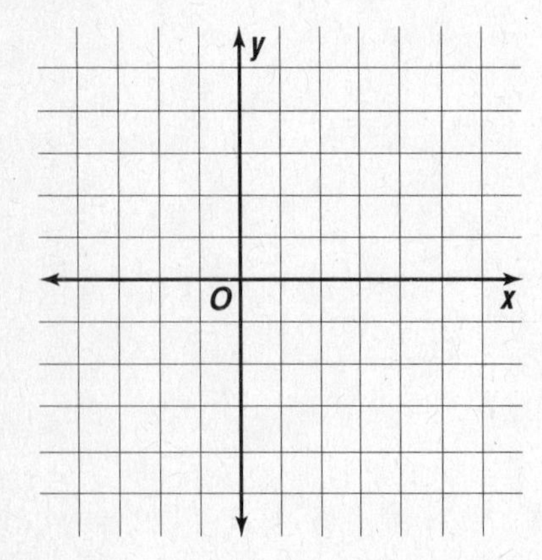

8. passes through $P(-1, 2)$ and is perpendicular to the line determined by $A(1, 4)$ and $B(2, 7)$

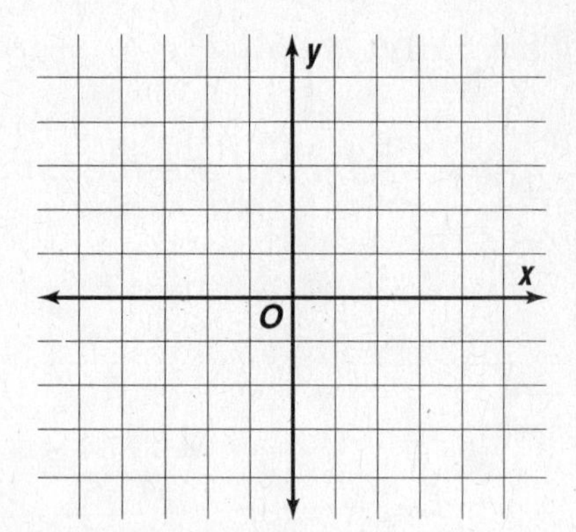

Determine the value of r so that a line through the points with the given coordinates has the given slope. Draw a sketch of each situation.

9. $(5, 3)$, $(r, 6)$; slope = 1

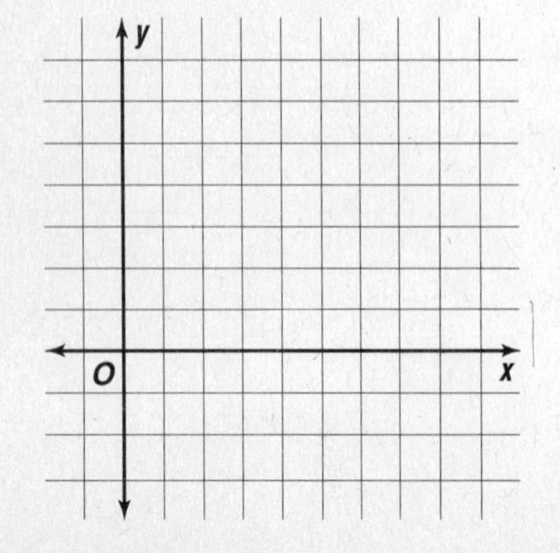

10. $(2, r)$, $(-2, 6)$; slope = $\frac{1}{2}$

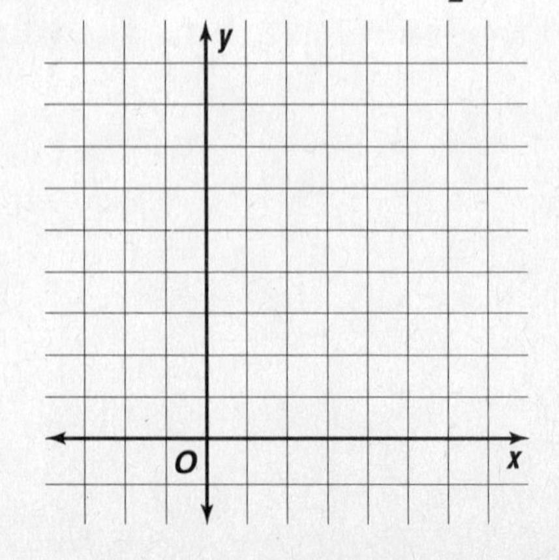

3–4

NAME ______________________________ DATE ____________

Practice

Student Edition
Pages 146–153

Proving Lines Parallel

For Exercises 1–6, find the value of x so that $\ell \parallel m$.

1.

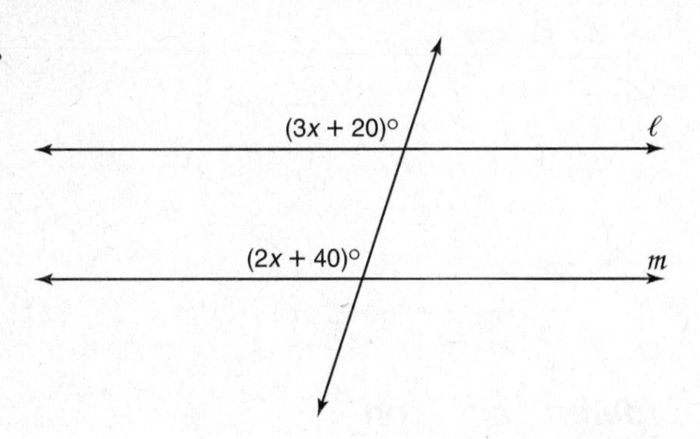

2.

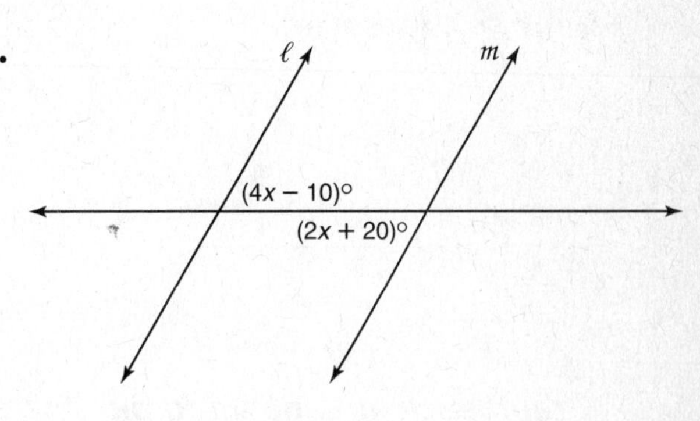

3.

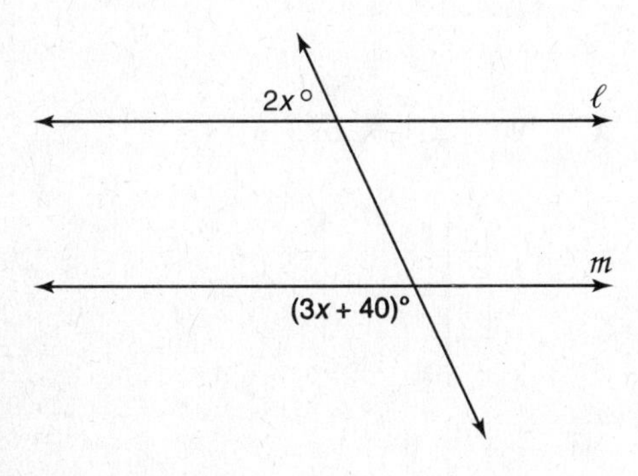

4.

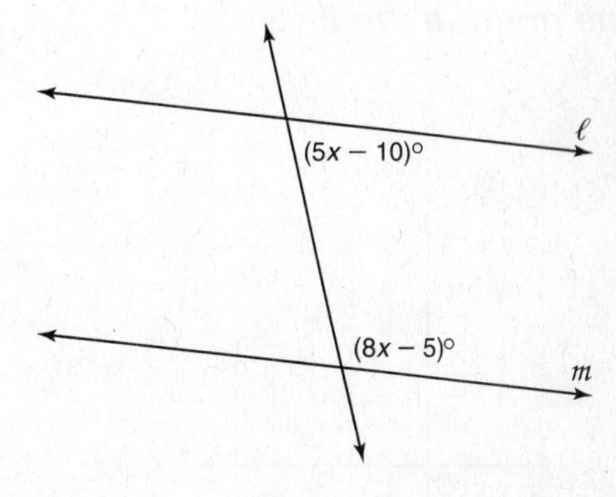

5.

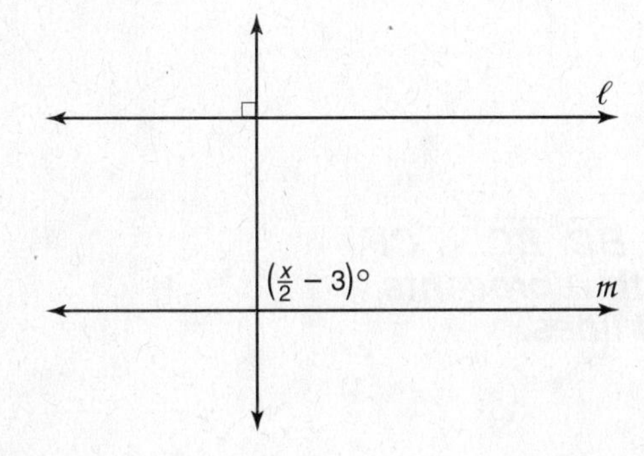

6.

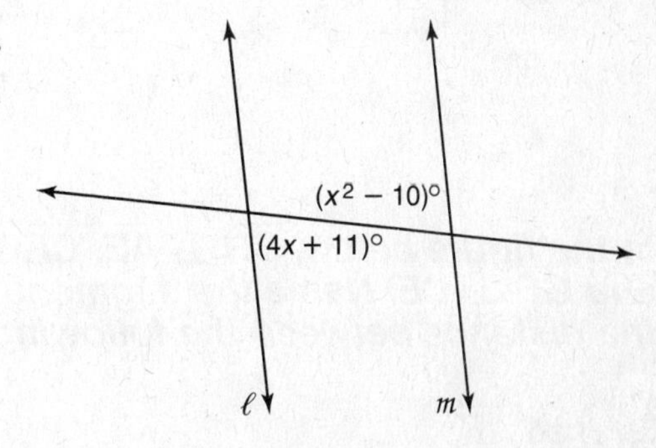

7. If $\ell \nparallel m$, can $x = 50$? Justify your answer.

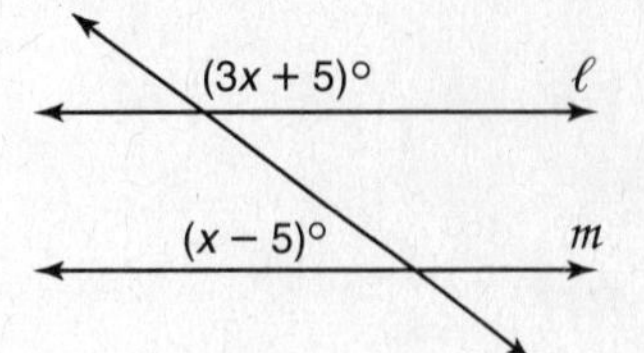

8. Find $m\angle 1$ for the figure at the right.

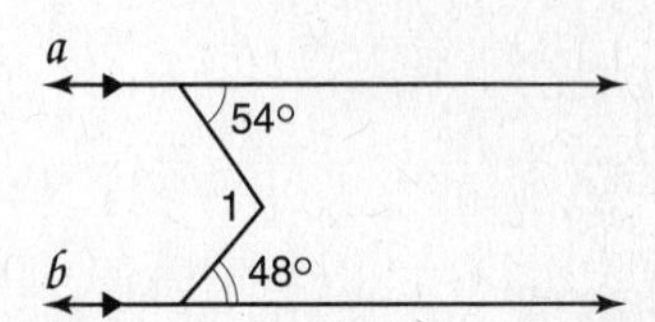

NAME ______________________ DATE ____________

Practice

Parallels and Distance

Draw the segment that represents the distance indicated.

1. P to $\overleftrightarrow{RS}$

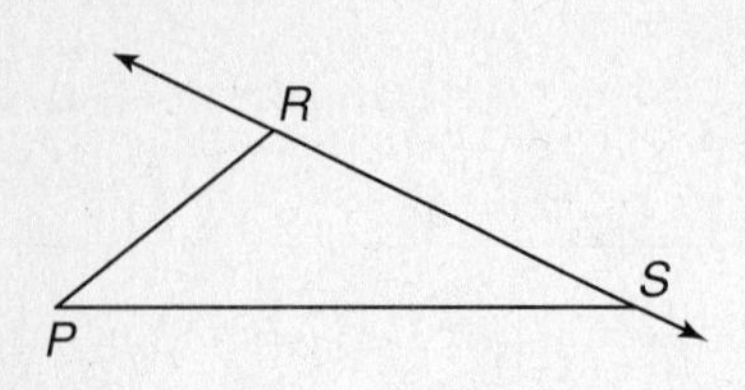

2. B to $\overline{AD}$

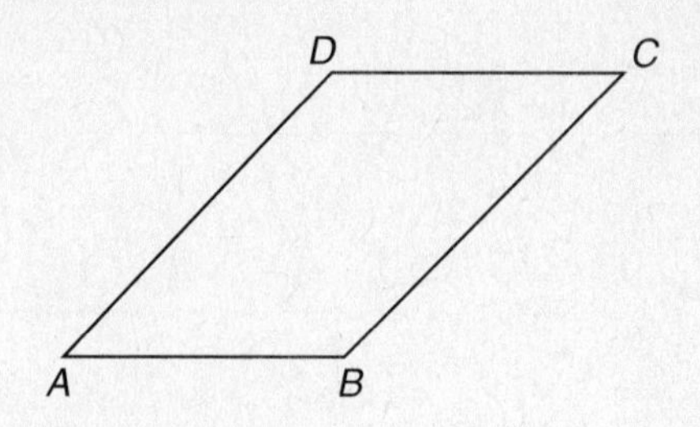

Graph each equation and plot the given ordered pair. Then construct a perpendicular segment and find the distance from the point to the line.

3. $y = x + 2$, $(2, -2)$

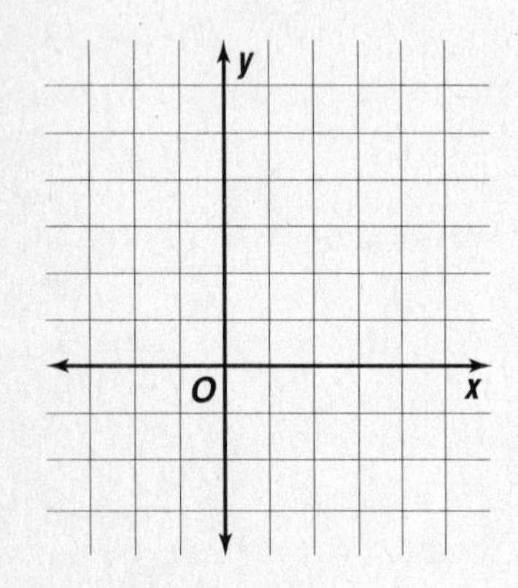

4. $x + y = 2$, $(3, 3)$

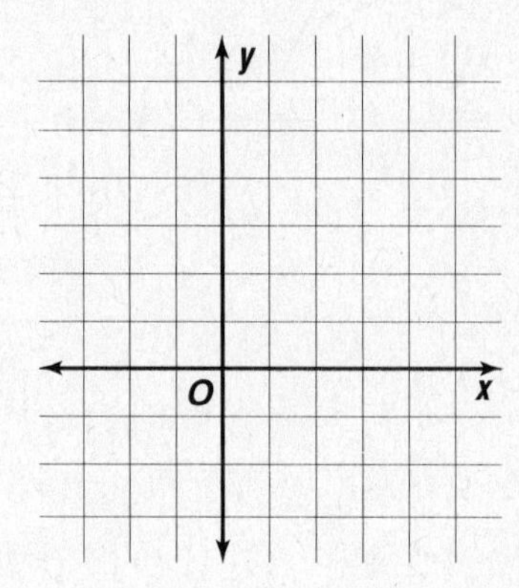

In the figure below, $\overline{BH} \perp \overline{AE}$, $\overline{CF} \perp \overline{AE}$, $\overline{BH} \perp \overline{BC}$, $\overline{BC} \perp \overline{CF}$, and $\overline{GD} \perp \overline{CE}$. Name the segment whose length represents the distance between the following points and lines.

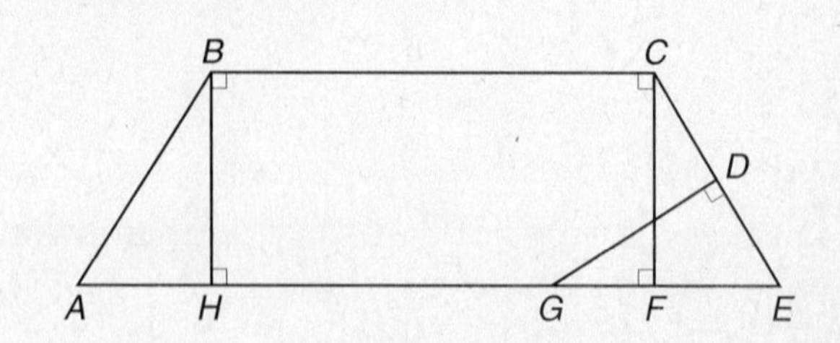

5. B to $\overline{AE}$

6. G to $\overline{CE}$

7. C to $\overline{BH}$

8. F to $\overline{BC}$

NAME ______________________ DATE __________

Practice

Integration: Non-Euclidean Geometry Spherical Geometry

Use a globe or world map to name the latitude and longitude of each city.

1. Bogota, Colombia

2. Cairo, Egypt

3. Melbourne, Australia

4. Fairbanks, Alaska

5. Nairobi, Kenya

6. London, England

7. Cape Town, South Africa

8. Julianehab, Greenland

9. La Paz, Bolivia

Use a globe or world map to name the city located near each set of coordinates.

10. 34°S, 59°W

11. 30°N, 90°W

12. 48°N, 123°W

13. 7°S, 107°E

14. 9°N, 38°E

15. 51°N, 14°E

16. 23°S, 46°W

17. 18°N, 99°W

18. 35°S, 173°E

For each property listed from plane Euclidean geometry, write a corresponding statement for non-Euclidean spherical geometry.

19. Two lines intersecting to form four right angles are perpendicular.

20. Through any two points in a plane there is a unique and infinite straight line.

21. An infinite number of lines can be drawn through a point in a plane.

Practice

Classifying Triangles

For Exercises 1–7, refer to the figure at the right. Triangle ABC is isosceles with AB > AC and AB > BC. Also, $\overleftrightarrow{XY} \parallel \overline{AB}$. Name each of the following.

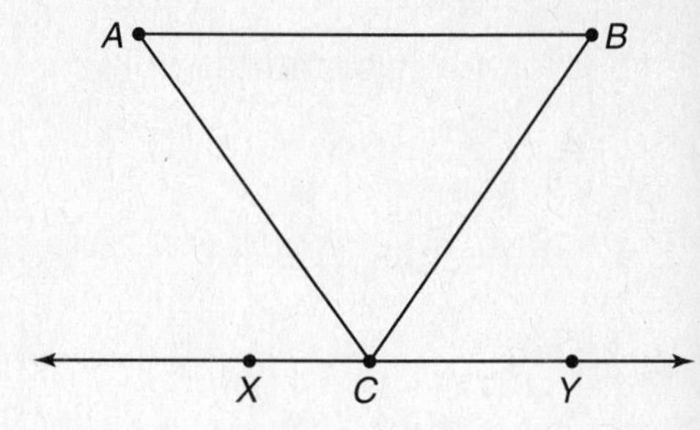

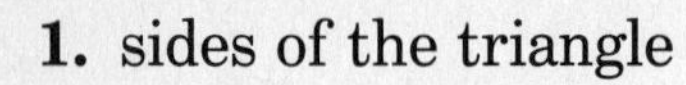

1. sides of the triangle

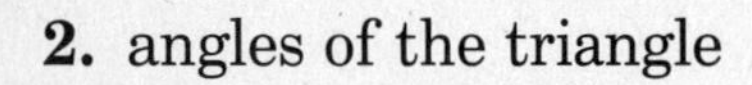

2. angles of the triangle

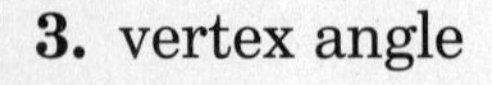

3. vertex angle

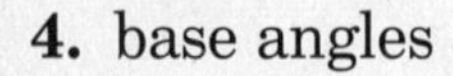

4. base angles

5. side opposite $\angle BCA$

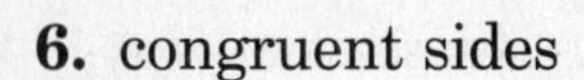

6. congruent sides

7. angle opposite $\overline{AC}$

Use a protractor and ruler to draw triangles using the given conditions. Classify each triangle by the measures of its angles and sides.

8. $\triangle BHE$, $BE = 1$ inch, $m\angle E = 60$, $HE = \frac{1}{2}$ inch

9. $\triangle QTR$, $m\angle T = 60$, $QT = TR = 4$ c

10. Find the measures of the legs of isosceles triangle ABC if $AB = 2x + 4$, $BC = 3x - 1$, $AC = x + 1$, and the perimeter of $\triangle ABC$ is 34 units.

NAME ____________________ DATE ________

Practice

Student Edition
Pages 189–195

Measuring Angles in Triangles

Find the value of x.

1.

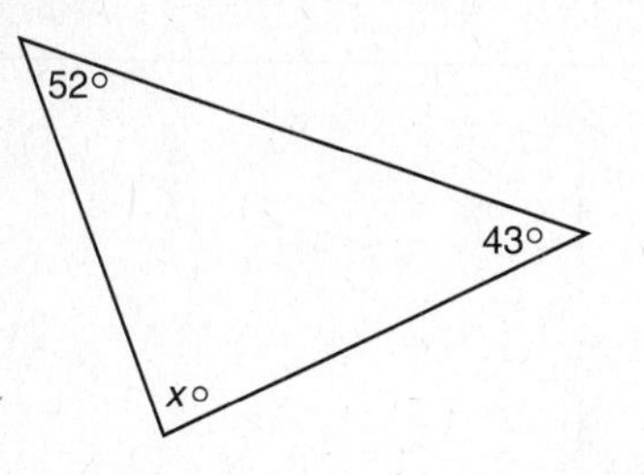

2.

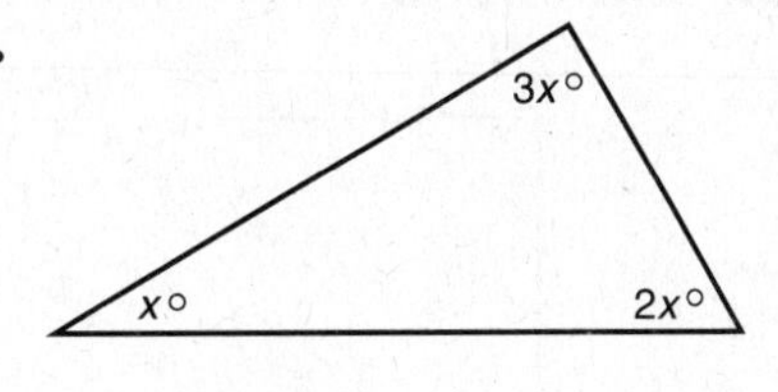

3.

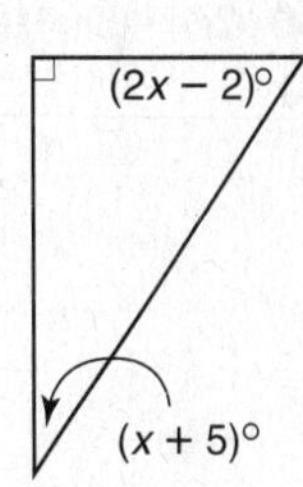

4.

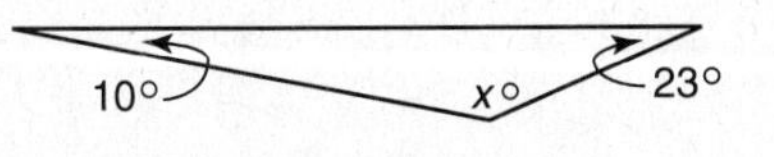

5.

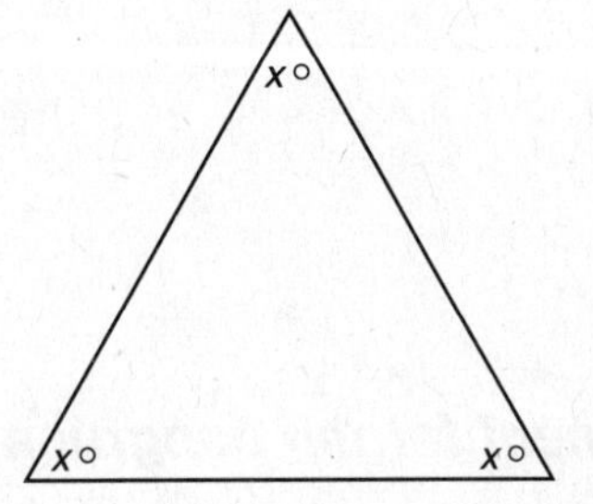

6.

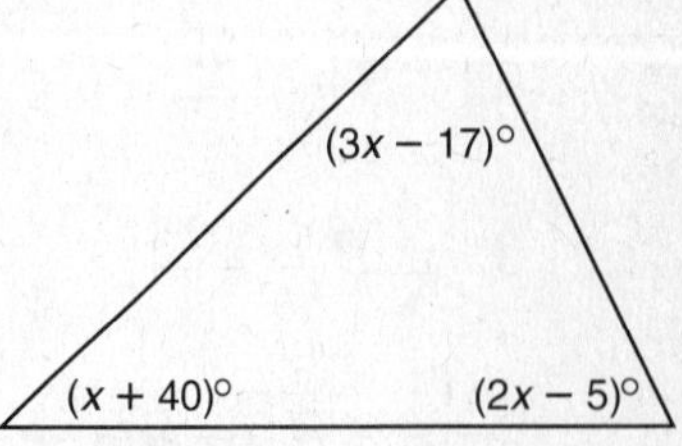

7.

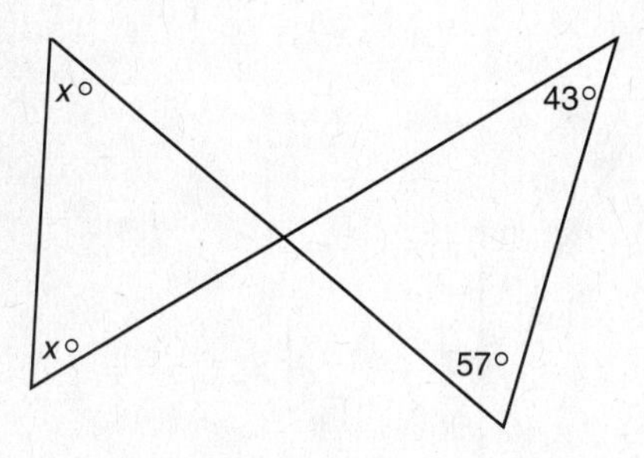

8.

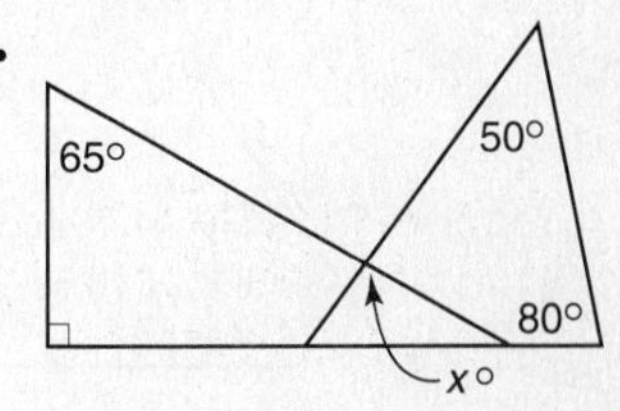

9.

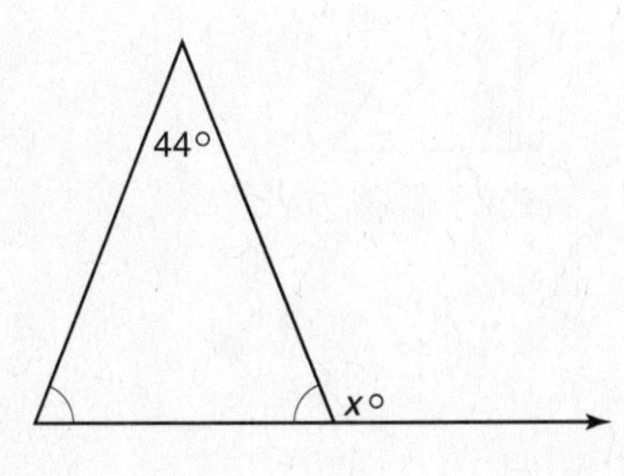

10.

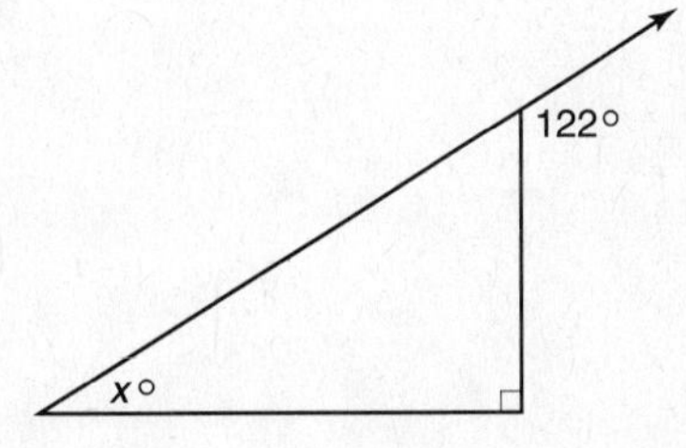

11.

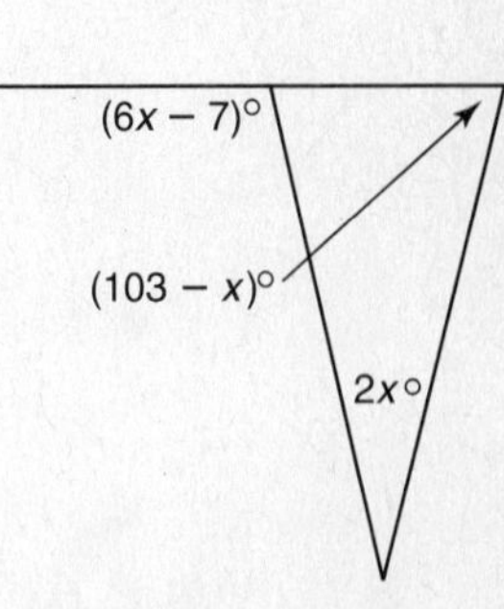

12.

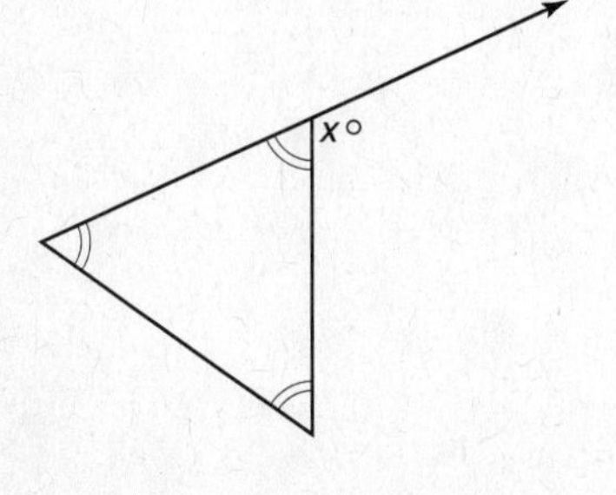

13.

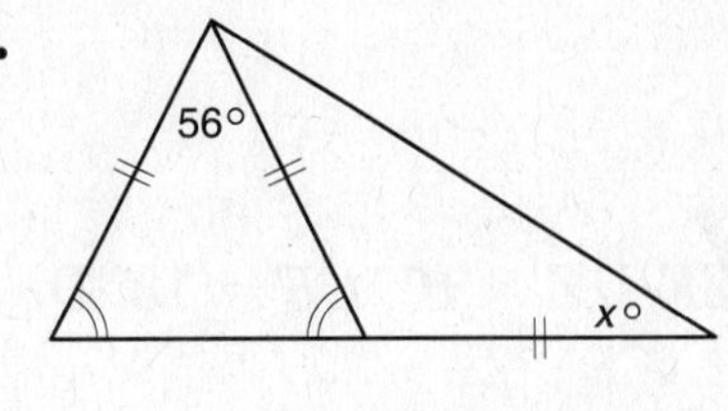

14.

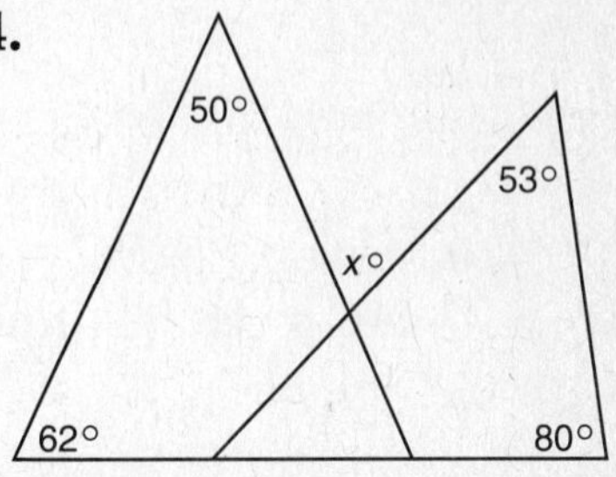

Exploring Congruent Triangles

Label the corresponding parts if $\triangle RST \cong \triangle ABC$. Use the figures to complete each statement.

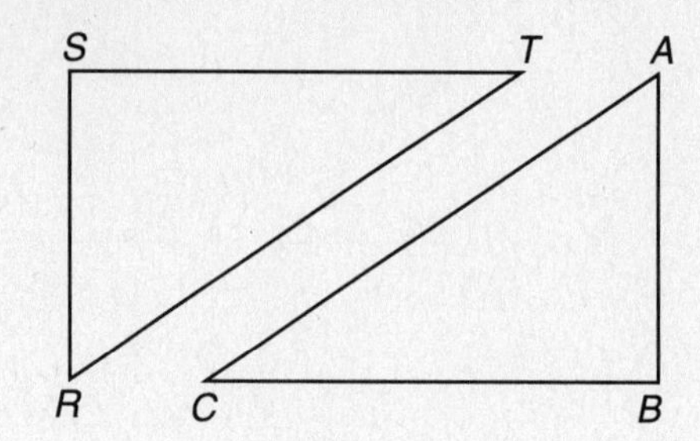

1. $\angle C \cong$ ___?___

2. $\angle R \cong$ ___?___

3. $\overline{AC} \cong$ ___?___

4. $\overline{ST} \cong$ ___?___

5. $\overline{RS} \cong$ ___?___

6. $\angle B \cong$ ___?___

Write a congruence statement for the congruent triangles in each diagram.

7.

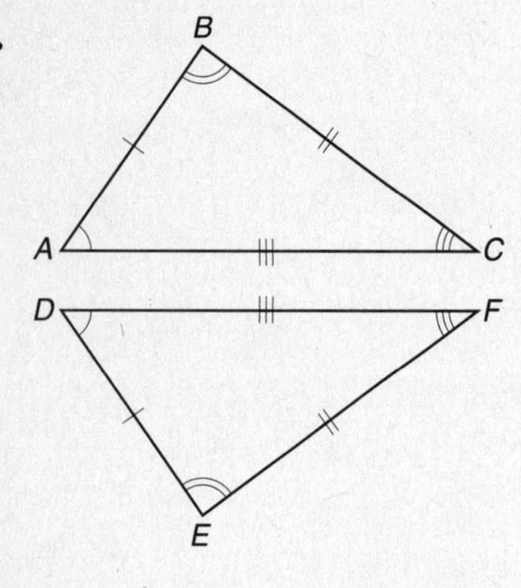

8.

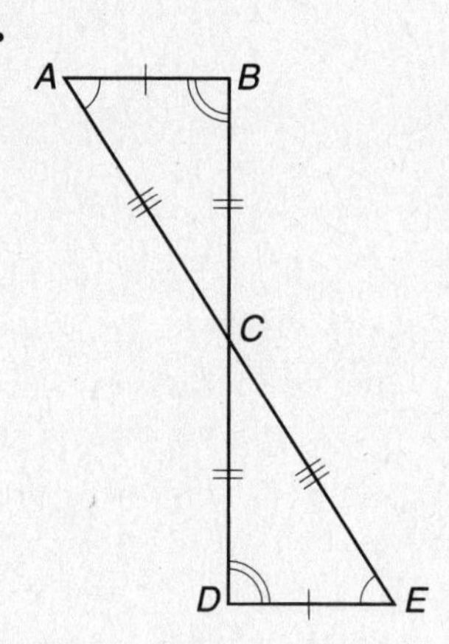

9. Given $\triangle ABC \cong \triangle DEF$, $AB = 15$, $BC = 20$, $AC = 25$, and $FE = 3x - 7$, find x.

10. Given $\triangle ABC \cong \triangle DEF$, $DE = 10$, $EF = 13$, $DF = 16$, and $AC = 4x - 8$, find x.

4–4

NAME________________________________ DATE ______________

Practice

Student Edition
Pages 206–213

Proving Triangles Congruent

For each figure, mark all congruent parts. Then complete the prove statement and identify the postulate that can be used to prove the triangles congruent.

1.

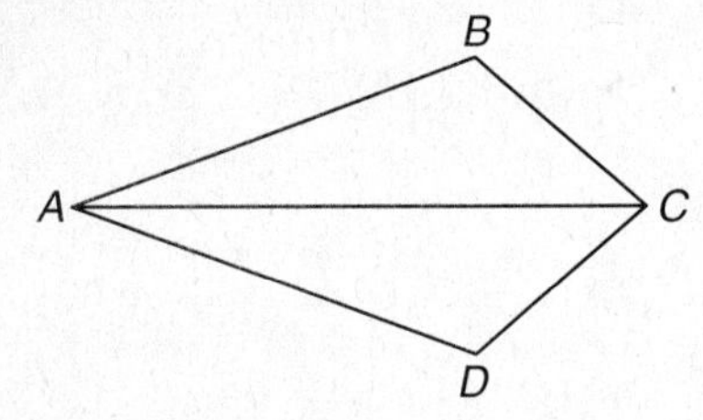

Given: $\overline{AB} \cong \overline{AD}$
$\overline{BC} \cong \overline{DC}$
Prove: $\triangle BCA \cong \underline{\ ?\ }$

2.

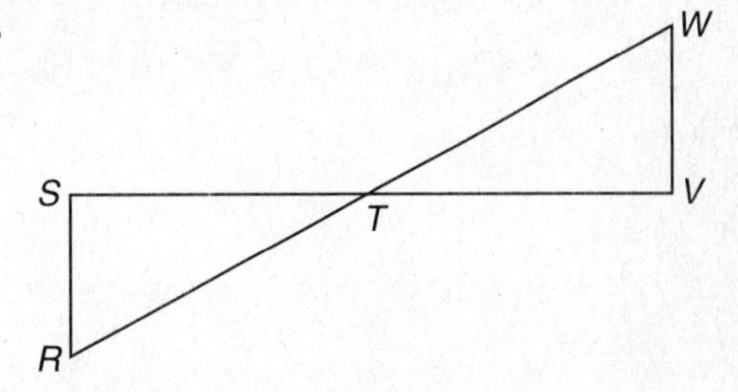

Given: $\angle S$ and $\angle V$ are right angles.
T bisects $\overline{SV}$.
Prove: $\triangle RST \cong \underline{\ ?\ }$

Write a two-column proof.

3. Given: $\overline{BD} \perp \overline{AC}$
D bisects $\overline{AC}$.
Prove: $\overline{AB} \cong \overline{CB}$
Proof:

Statements	Reasons

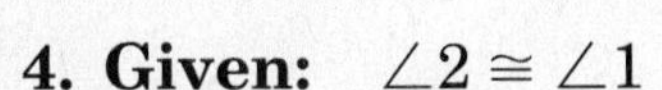

4. Given: $\angle 2 \cong \angle 1$
$\angle 4 \cong \angle 5$
Prove: $\overline{BC} \cong \overline{DC}$
Proof:

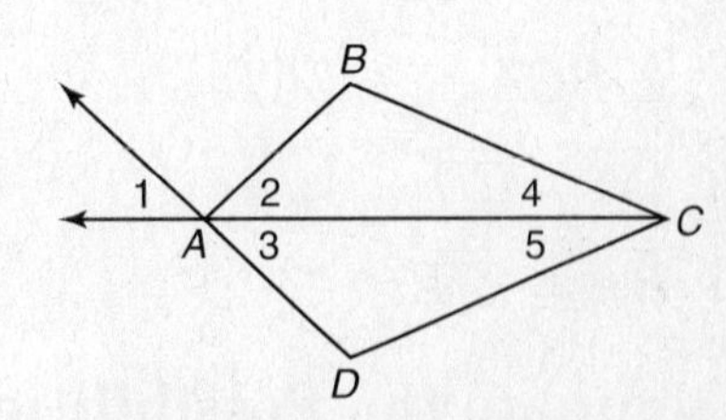

Statements	Reasons

NAME ______________________ DATE ________

Practice

More Congruent Triangles

Draw and label triangles MNO and XYZ. Indicate the additional pairs of corresponding parts that would have to be proved congruent in order to use the given postulate or theorem to prove the triangles congruent.

1. $\angle N \cong \angle Y$ and $\overline{NO} \cong \overline{YZ}$ by ASA

2. $\angle O \cong \angle Z$ and $\angle M \cong \angle X$ by AAS

3. $\angle O \cong \angle Z$ and $\overline{MO} \cong \overline{XZ}$ by AAS

4. $\angle N \cong \angle Y$ and $\angle M \cong \angle X$ by ASA

5. The statements in the following proof are *not* in logical order. Rearrange them in a correct sequence and give the reasons.

Given: $\angle A \cong \angle D$
$\overline{AB} \cong \overline{DE}$

Prove: $\overline{CA} \cong \overline{CD}$

Proof:

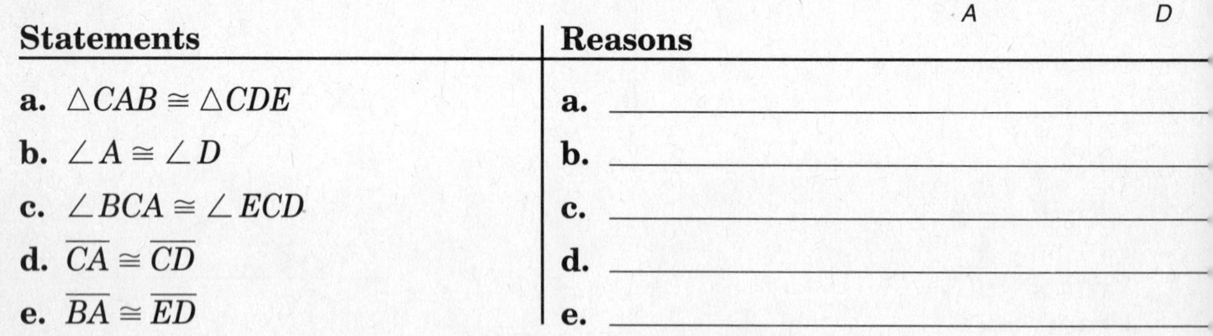

Statements	Reasons
a. $\triangle CAB \cong \triangle CDE$	**a.** ______
b. $\angle A \cong \angle D$	**b.** ______
c. $\angle BCA \cong \angle ECD$	**c.** ______
d. $\overline{CA} \cong \overline{CD}$	**d.** ______
e. $\overline{BA} \cong \overline{ED}$	**e.** ______

6. Eliminate the Possibilities Barky, Spot, and Tiger are dogs. One is a black labrador retriever, one is a multi-colored collie, and one is a spotted dalmatian. One is owned by a doctor, one is owned by a lawyer, and one is owned by an insurance salesperson. The lawyer's dog does not have spots. The doctor's dog has more than one color. The insurance salesperson's dog is solid in color. Who owns which dog?

4–6

NAME ______________________ DATE ____________

Practice

Student Edition
Pages 222–228

Analyzing Isosceles Triangles

Find the value of x.

1.

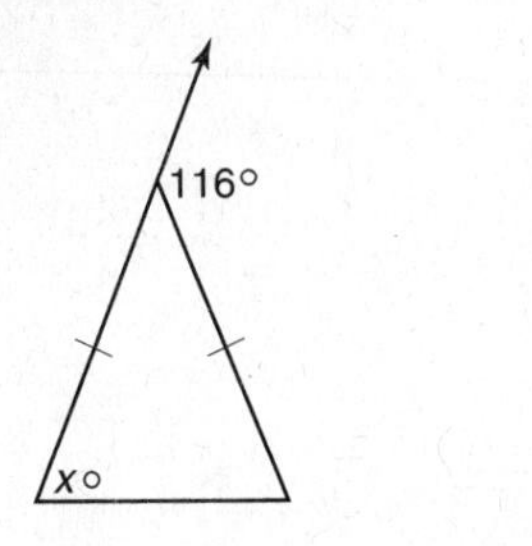

2.

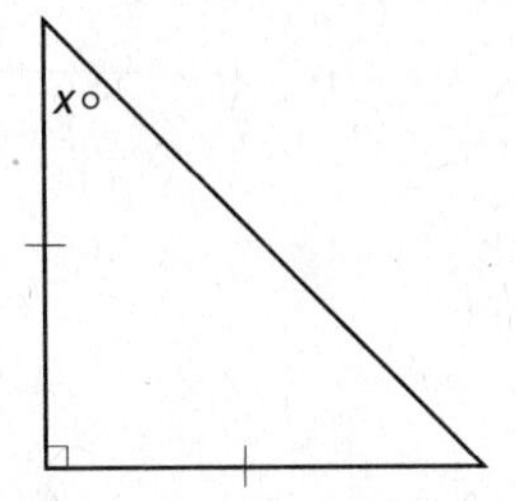

3.

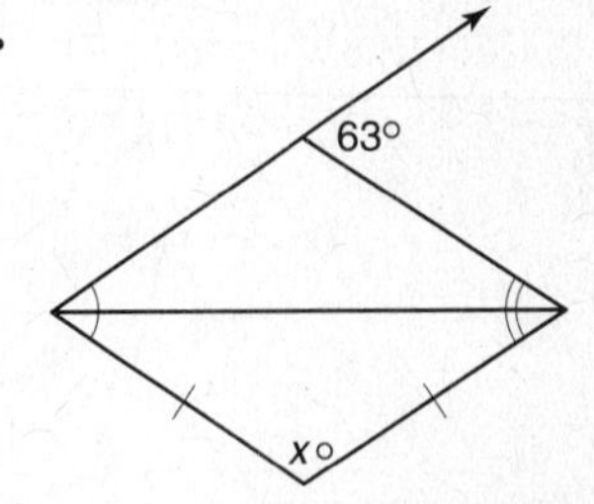

4.

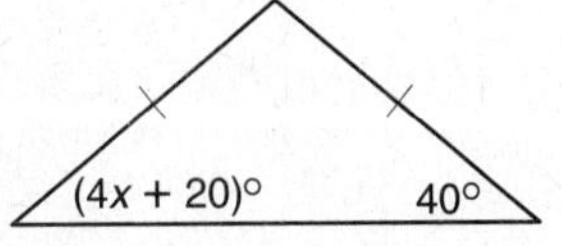

5.

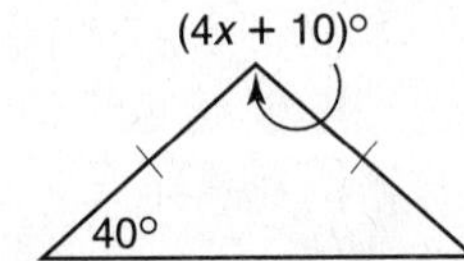

6.

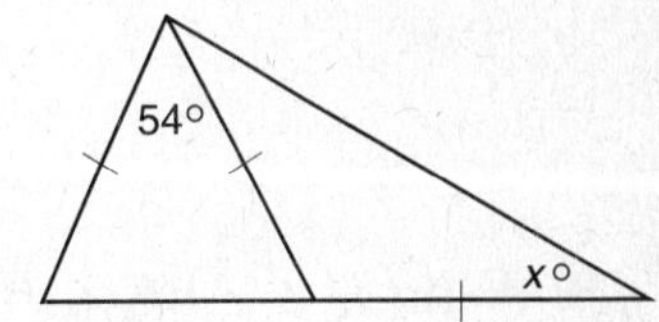

Write a two-column proof.

7. Given: $\angle 3 \cong \angle 4$
Prove: $\overline{AB} \cong \overline{BC}$
Proof:

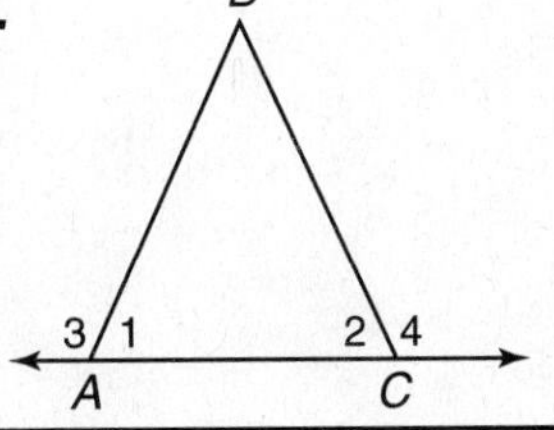

Statements	Reasons

8. Given: $\overline{AR} \cong \overline{AQ}$
$\overline{RS} \cong \overline{QT}$
Prove: $\overline{AS} \cong \overline{AT}$
Proof:

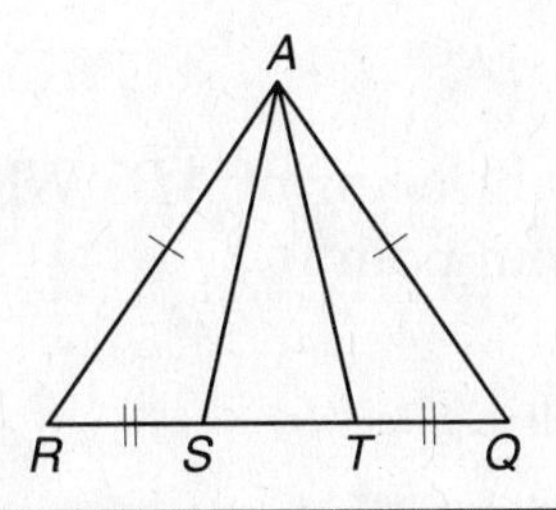

Statements	Reasons

Practice

Special Segments in Triangles

1. Find AB if $\overline{BD}$ is a median of $\triangle ABC$.

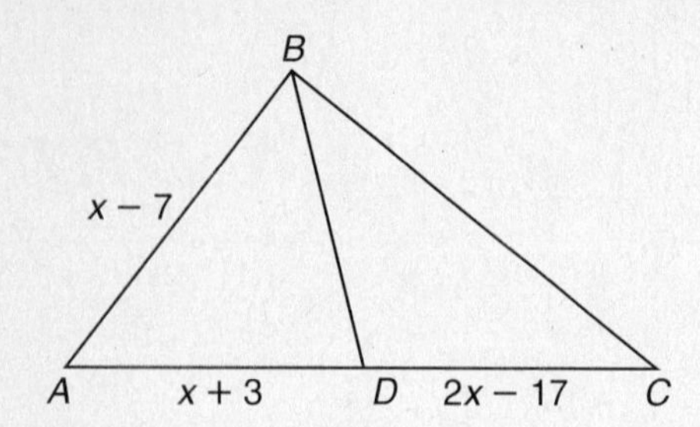

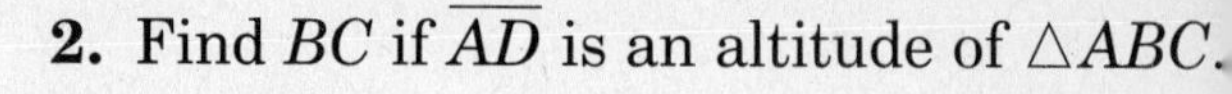

2. Find BC if $\overline{AD}$ is an altitude of $\triangle ABC$.

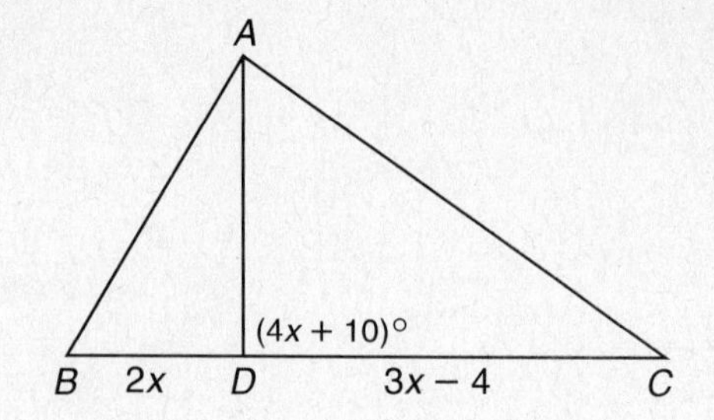

3. Find $m\angle ABC$ if $\overline{BD}$ is an angle bisector of $\triangle ABC$.

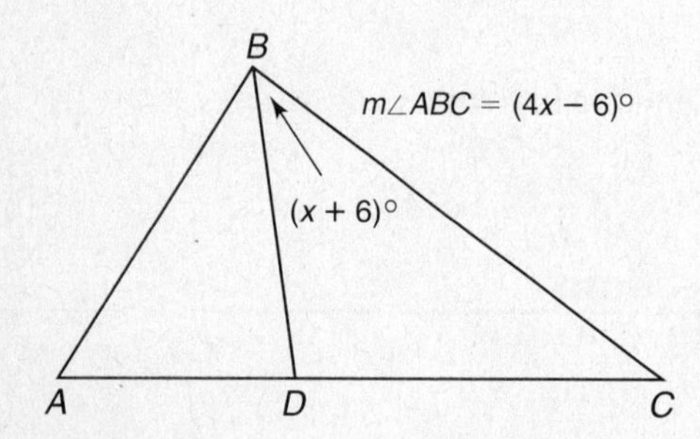

In Exercises 4–6, A(2, 5), B(12, −1), and C(−6, 8) are the vertices of △ ABC.

4. What are the coordinates of K if $\overline{CK}$ is a median of $\triangle ABC$?

5. What is the slope of the perpendicular bisector of $\overline{AB}$? What is the slope of $\overline{CL}$ if $\overline{CL}$ is the altitude from point C?

6. Point N on $\overleftrightarrow{BC}$ has coordinates $\left(\frac{8}{5}, \frac{21}{5}\right)$. Is $\overline{NA}$ an altitude of $\triangle ABC$? Explain your answer.

Practice

Right Triangles

For each figure, find the values of x and y so that $\triangle DEF \cong \triangle PQR$ by the indicated theorem or postulate.

1. HA

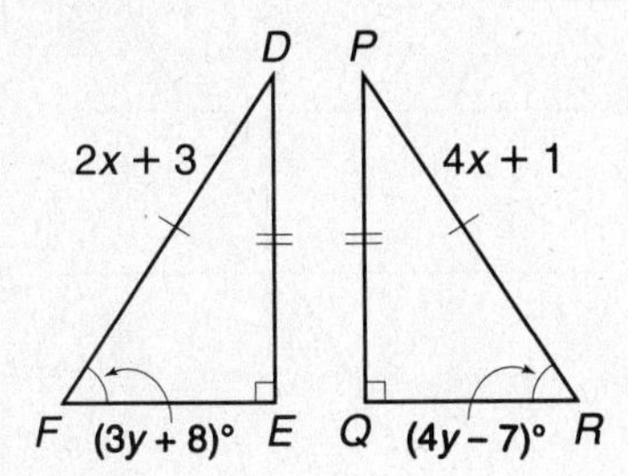

2. LL

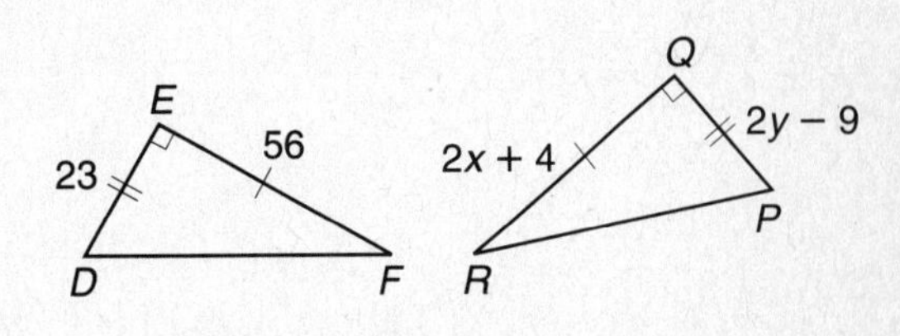

3. LA

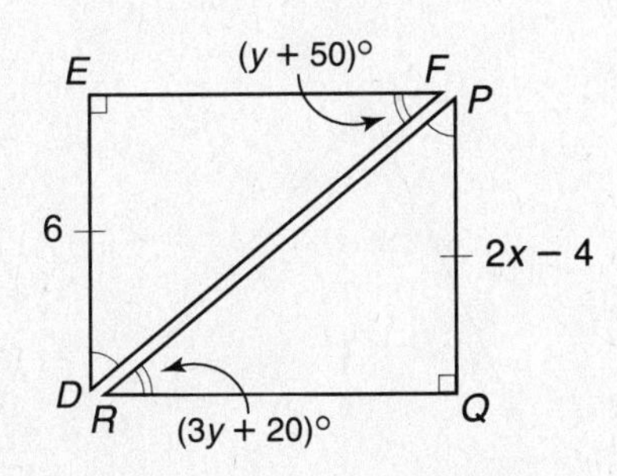

4. HL

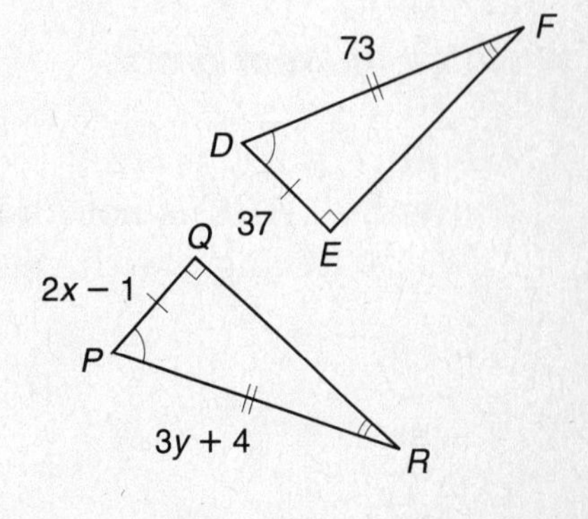

5. Write a two-column proof.
Given: $\overline{AB}$ bisects $\angle DAC$
$\angle C$ and $\angle D$ are right angles.
Prove: $\overline{BC} \cong \overline{BD}$
Proof:

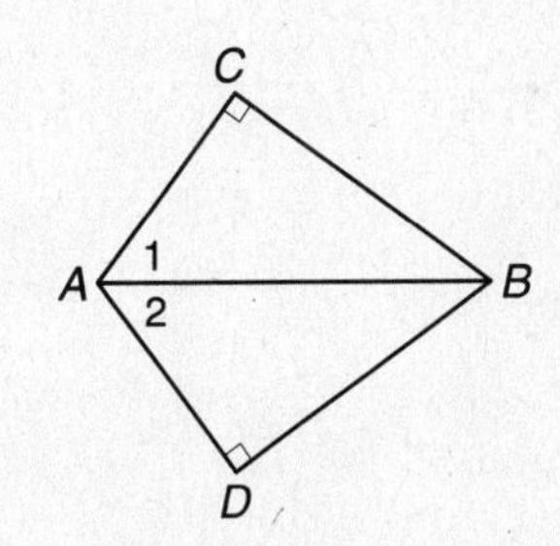

Statements	**Reasons**

Indirect Proof and Inequalities

Use the figure at the right to complete each statement with either < or >.

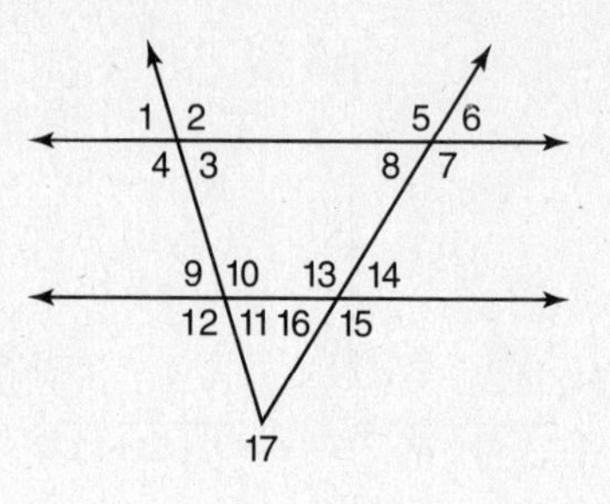

1. $\angle 4$ __?__ $\angle 8$

2. $\angle 13$ __?__ $\angle 11$

3. $\angle 17$ __?__ $\angle 8$

4. If $m\angle 15 = m\angle 7$ then $m\angle 11$ __?__ $m\angle 7$.

Write an indirect proof.

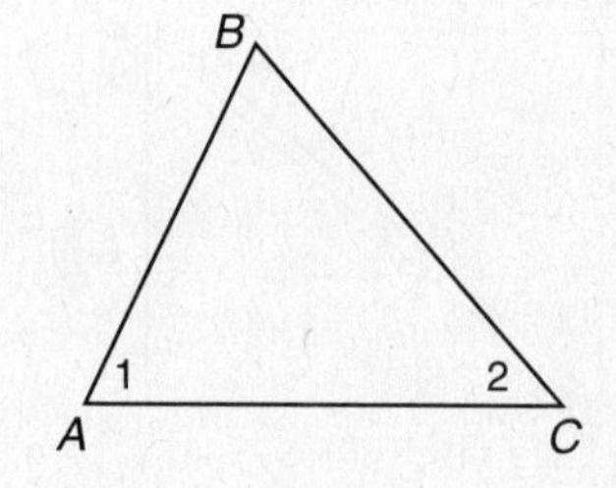

5. Given: $m\angle 1 \neq m\angle 2$
Prove: $\triangle ABC$ is not an isosceles triangle with vertex B.

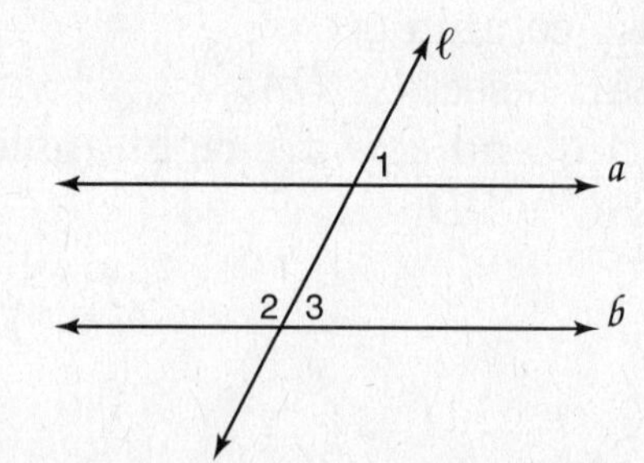

6. Given: $m\angle 1 + m\angle 2 \neq 180$
Prove: $a \nparallel b$

7. Work Backward Joe spent half of the money in his wallet on a table for his computer printer. He then spent $8.12 for a printer ribbon. Then he spent half of what he had left on supplies for his office. He then had $14.50 remaining. How much money did Joe have to start with?

NAME ______________________________ DATE ____________

Practice

Inequalities for Sides and Angles of a Triangle

Refer to the figure on the right for Exercises 1–4.

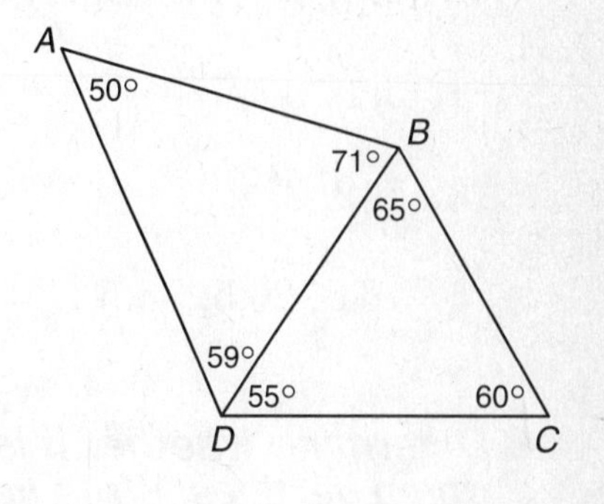

1. Name the shortest and the longest segments in $\triangle BCD$.

2. Name the shortest and the longest segments in $\triangle ABD$.

3. Find the shortest segment in the figure.

4. How many of the segments in the figure are longer than $\overline{BD}$?

5. List the angles in order from least to greatest.

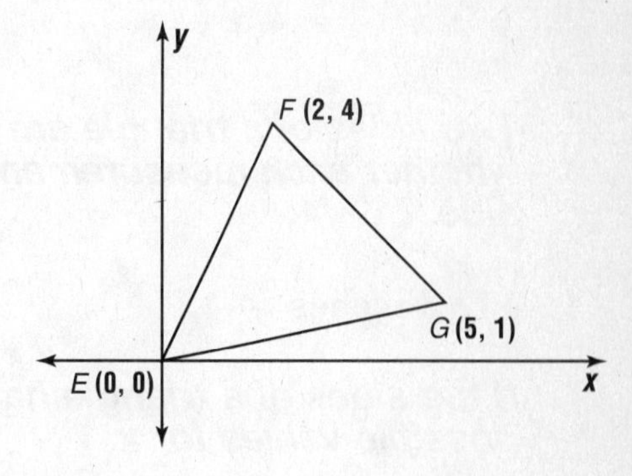

6. List the sides of $\triangle MNO$ in order from longest to shortest if $m\angle M = 4x + 20$, $m\angle N = 2x + 10$, and $m\angle O = 3x - 20$.

7. List the angles of $\triangle KLM$ in order from least to greatest if $KL = x - 4$, $LM = x + 4$, $KM = 2x - 1$, and the perimeter of $\triangle KLM$ is 27.

Practice

The Triangle Inequality

Determine whether it is possible to draw a triangle with sides of the given measure. Write yes or no.

1. 3, 3, 3

2. 2, 3, 4

3. 1, 2, 3

4. 8.9, 9.3, 18.3

5. 16.5, 20.5, 38.5

6. 19, 19, 0.5

Determine whether it is possible to have a triangle with the given vertices. Write yes or no, and explain your answer.

7. $A(-2, -2)$, $B(-1, 1)$, $C(1, 4)$

8. $A(-4, 2)$, $B(-2, 1)$, $C(2, -1)$

9. $A(2, 5)$, $B(-3, 5)$, $C(6, -1)$

10. $A(3, -6)$, $B(1, 2)$, $C(-2, 10)$

Two sides of a triangle are 21 and 24 inches long. Determine whether each measurement can be the length of the third side.

11. 3 inches

12. 40 inches

13. 56 inches

If the sides of a triangle have the following lengths, find all possible values for x.

14. $AB = 2x + 5$, $BC = 3x - 2$, $AC = 4x - 8$

15. $PQ = 3x$, $QR = 4x - 7$, $PR = 2x + 9$

NAME ________________________ DATE ____________

Practice

Student Edition
Pages 273–279

Inequalities Involving Two Triangles

Refer to each figure to write an inequality relating the given pair of angle measures.

1. $m\angle PRQ$, $m\angle PRS$

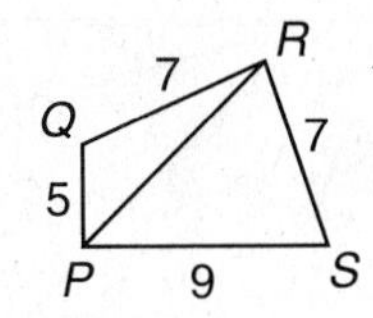

2. $m\angle ABE$, $m\angle DBC$

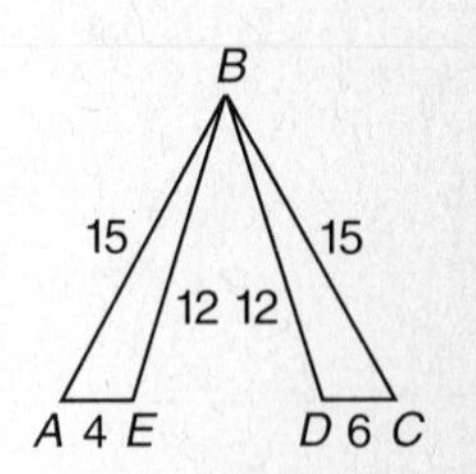

Write an inequality or pair of inequalities to describe the possible values of x.

3.

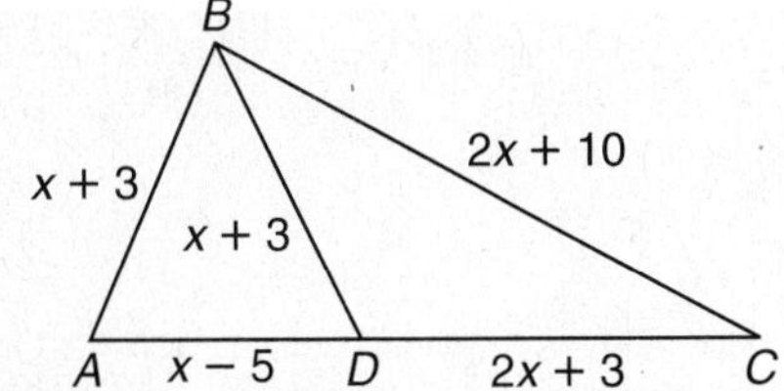

4.

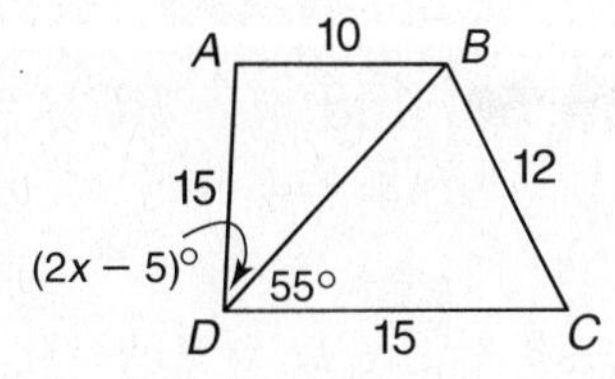

Write a two-column proof.

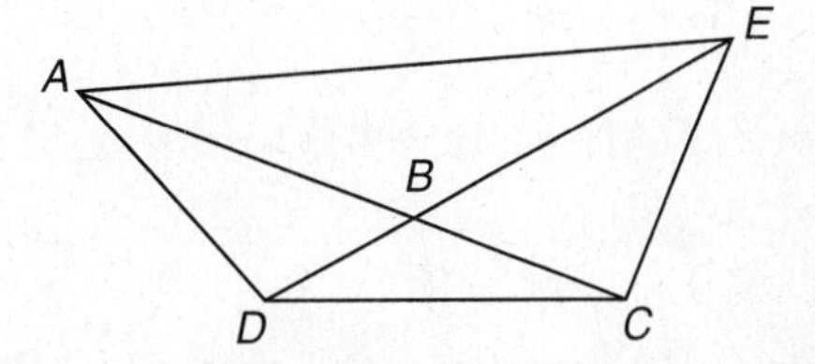

5. Given: $\overline{AD} \cong \overline{EC}$
$m\angle ADC > m\angle ECD$
Prove: $AC > ED$
Proof:

Statements	Reasons

6. Given: D is the midpoint of $\overline{AC}$.
$BC > AB$
Prove: $m\angle 1 > m\angle 2$
Proof:

Statements	Reasons

Parallelograms

The figure at the right is a parallelogram. Use this figure and the information given to solve each problem.

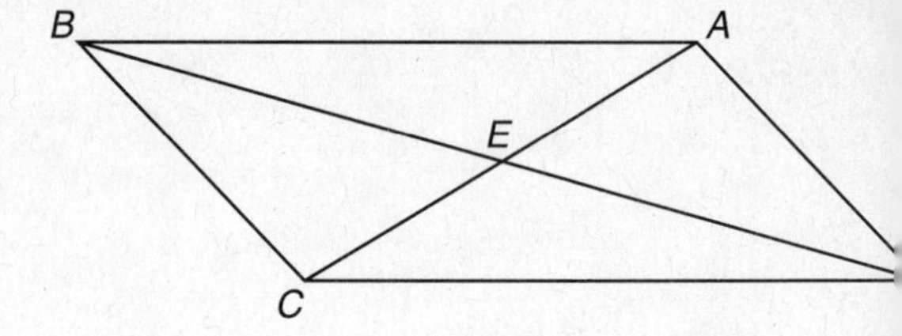

1. If $m\angle BCD = 125$, find $m\angle BAD$.

2. If $m\angle BAC = 45$, find $m\angle ACD$.

3. If $m\angle BEA = 135$, find $m\angle AED$.

4. If $m\angle ABC = 50$, find $m\angle BCD$.

5. If $AB = 5x - 3$ and $CD = 2x + 9$, find AB.

6. If $m\angle DAB = 2x - 10$ and $m\angle ADC = 3x$, find $m\angle DAB$.

7. If $m\angle BAD = 3x - 12$ and $m\angle BCD = x + 40$, find $m\angle BAD$.

8. Write a two-column proof.
Given: $ABCD$ is a parallelogram, $\overline{BE} \cong \overline{AD}$
Prove: $\angle 1 \cong \angle C$

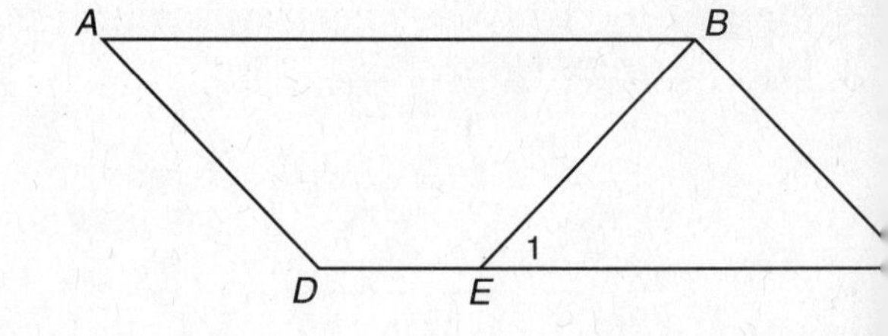

Proof:

Statements	Reasons

NAME ______________________ DATE ________

Practice

Student Edition
Pages 298–304

Tests for Parallelograms

Find the values of x and y that insure each quadrilateral is a parallelogram.

1.

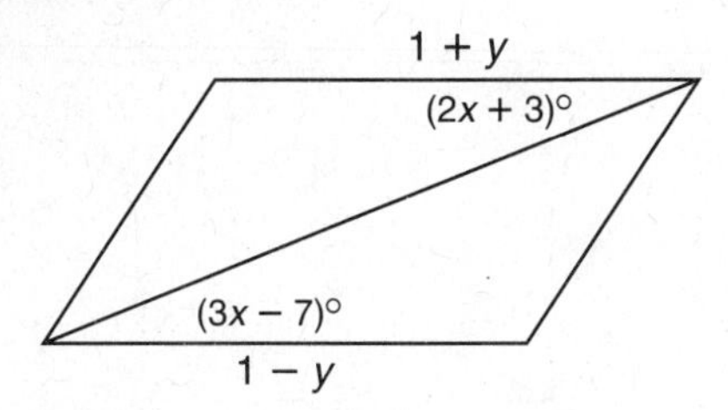

2.

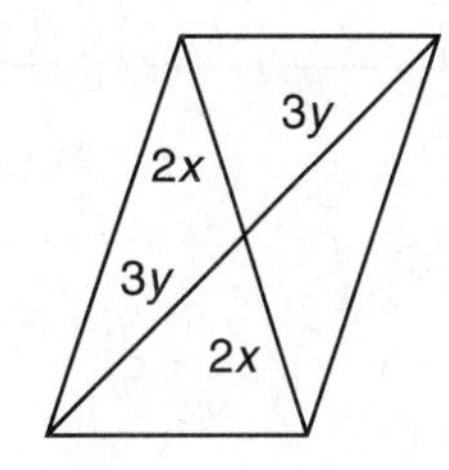

3. Refer to the figure at the right. $\overline{YZ}$ bisects $\angle XYK$ and $\overline{LK}$ bisects $\angle ZLM$. Also, $\angle 1 \cong \angle 2$.
Is *YZLK* a parallelogram? Explain.

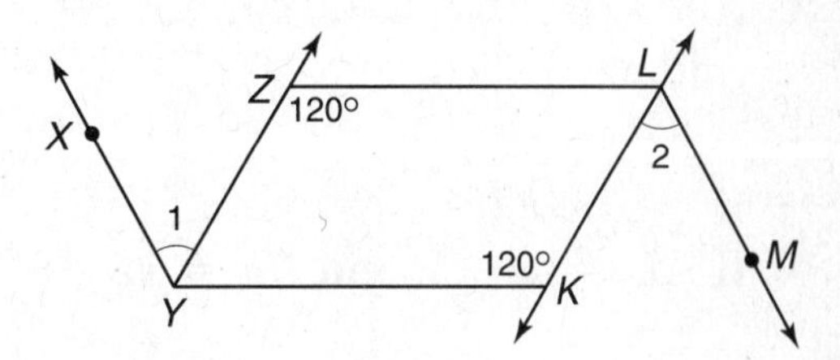

4. Refer to the figure at the right. $\overline{AC} \cong \overline{WV}$ and $\overline{BD} \cong \overline{WV}$. Also, $\overline{AB} \cong \overline{XY}$ and $\overline{CD} \cong \overline{XY}$. Is ABDC a parallelogram? Explain.

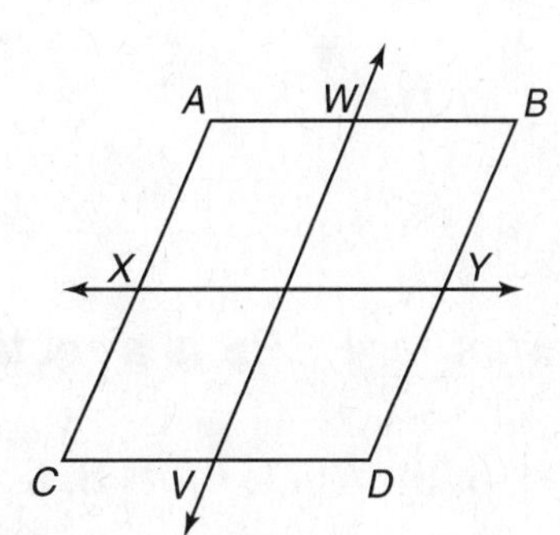

Determine whether quadrilateral ABCD with the given vertices is a parallelogram. Explain.

5. $A(2, 5)$, $B(5, 9)$, $C(6, 3)$, $D(3, -1)$

6. $A(-1, 6)$, $B(2, -3)$, $C(5, 9)$, $D(2, 7)$

7. Identify Subgoals Identify the subgoals you would need to accomplish to complete the proof.

Given: $\overline{YN} \perp \overline{XZ}$, $\overline{ZM} \perp \overline{XY}$
$\overline{XZ} \cong \overline{XY}$
$\overline{XM} \cong \overline{XN}$

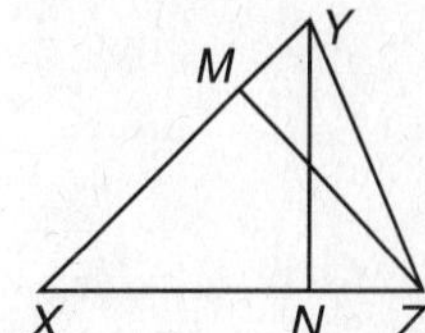

Prove: $\angle XMZ \cong \angle XNY$

Practice

Rectangles

Use rectangle ABCD and the given information to solve each problem.

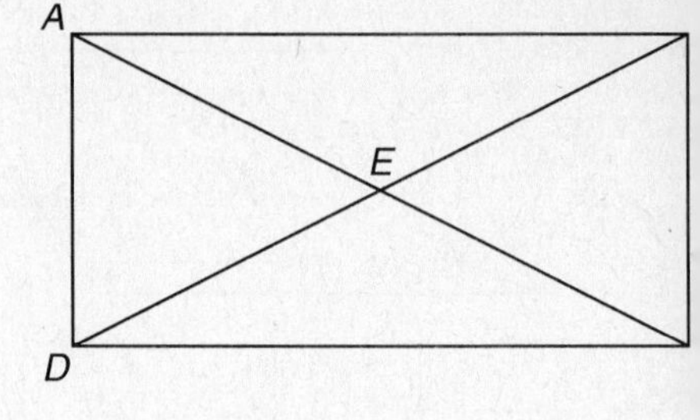

1. If $AC = 4x - 60$ and $BD = 30 - x$, find BD.

2. If $AC = 4x - 60$ and $AE = x + 5$, find EC.

3. If $m\angle BAC = 4x + 5$ and $m\angle CAD = 5x - 14$, find $m\angle CAD$.

4. If $AE = 2x + 3$ and $BE = 12 - x$, find BD.

5. If $m\angle BAC = 3x + 5$ and $m\angle ACD = 40 - 2x$. Find $m\angle AED$.

Determine whether PQRS is a rectangle. Justify your answer.

6. $P(2, 3)$, $Q(5, 9)$, $R(11, 6)$, $S(8, 0)$

7. $P(-1, 4)$, $Q(3, 6)$, $R(9, -3)$, $S(5, -5)$

8. $P(1, 3)$, $Q(4, 7)$, $R(6, 2)$, $S(2, 4)$

9. $P(-1, -3)$, $Q(-4, 6)$, $R(8, 10)$, $S(11, 1)$

10. $P(-1, -2)$, $Q(5, 2)$, $R(13, -10)$, $S(7, -14)$

NAME ______________________________ DATE ____________

Practice

Squares and Rhombi

Use square ABCD and the given information to find each value.

1. If $m\angle AEB = 3x$, find x.

2. If $m\angle BAC = 9x$, find x.

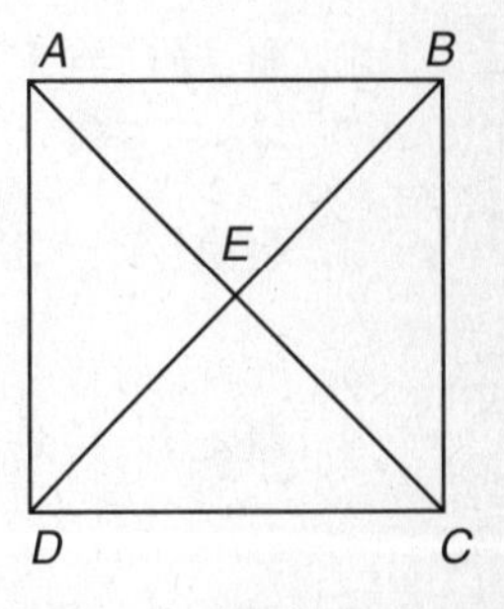

3. If $AB = 2x + 4$ and $CD = 3x - 5$, find BC.

4. If $m\angle DAC = y$ and $m\angle BAC = 3x$, find x and y.

5. If $AB = x^2 - 15$ and $BC = 2x$, find x.

Use rhombus ABCD and the given information to find each measure.

6. $m\angle BCE$

7. $m\angle BEC$

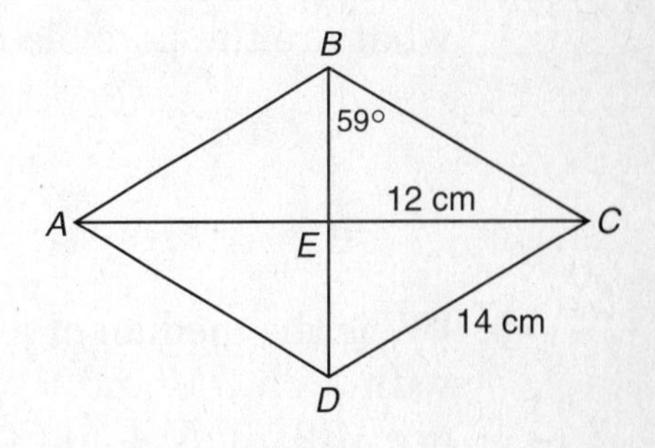

8. AC

9. $m\angle ABD$

10. AD

Determine whether EFGH is a parallelogram, a rectangle, a rhombus, or a square for each set of vertices. List all that apply.

11. $E(0, -3)$, $F(-3, 0)$, $G(0, 3)$, $H(3, 0)$

12. $E(2, 1)$, $F(3, 4)$, $G(7, 2)$, $H(6, -1)$

Practice

Trapezoids

MATH is an isosceles trapezoid with bases $\overline{MA}$ and $\overline{TH}$. Use the given information to solve each problem.

1. If $MA = 34$ and $HT = 20$, find CD.

2. If $HT = 17.6$ and $CD = 28.6$ find MA.

3. If $MA = 23.9$ and $CD = 16.4$, find HT.

4. If $CD = x + 12$ and $MA + HT = 4x + 3$, find x.

5. If $m\angle TAM = 63$, find $m\angle HMA$.

6. If $m\angle HCD = 52$, find $m\angle TDC$.

7. If $m\angle DCM = 2x$, find $m\angle CMA$ in terms of x.

8. If the measure of the median of an isosceles trapezoid is 5.5, what are the possible integral measures for the bases?

9. $\overline{VW}$ is the median of a trapezoid that has bases $\overline{MN}$ and $\overline{PO}$, with V on $\overline{OM}$ and W on $\overline{PN}$. If the vertices of the trapezoid are $M(2, 6)$, $N(4, 6)$ $P(10, 0)$, and $O(0, 0)$, find the coordinates of V and W.

10. $\overline{VW}$ is the median of a trapezoid that has bases $\overline{MN}$ and $\overline{PO}$, with V on $\overline{PM}$ and W on $\overline{ON}$. If four of the points are $M(5, 10)$, $N(9, 10)$, $V(3, 7)$, and $W(11, 7)$, find the coordinates of P and O.

NAME ______________________ DATE ____________

Practice

Student Edition
Pages 338–345

Integration: Algebra
Using Proportions

Solve each proportion using cross products.

1. $\frac{3}{5} = \frac{x}{15}$

2. $\frac{20 - x}{x} = \frac{6}{4}$

3. $\frac{x + 1}{5} = \frac{x - 1}{2}$

4. $\frac{x}{x - 3} = \frac{x + 4}{x}$

5. $\frac{x + 1}{6} = \frac{x - 1}{x}$

6. $\frac{1}{x} = \frac{6}{x + 9}$

7. $\frac{x}{x + 8} = \frac{2}{3}$

8. $\frac{4}{12} = \frac{x + 2}{2x + 5}$

In the figure at the right, $\frac{AC}{CD} = \frac{CE}{CB}$. Use proportions to complete the table.

	AC	BC	AB	CE	ED	DC
9.	10	4		8		
10.	12			10		9

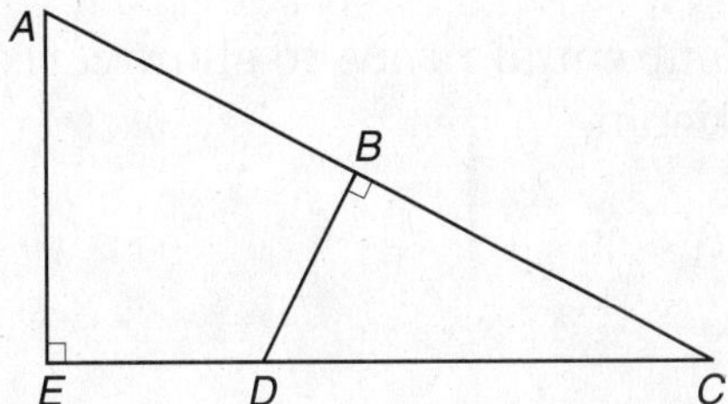

Use a proportion to solve each problem.

11. The ratio of seniors to juniors in the Math Club is 2:3. If there are 21 juniors, how many seniors are in the club?

12. A 15-foot building casts a 9-foot shadow. How tall is a building that casts a 30-foot shadow at the same time?

13. A photo that is 3 inches wide and 5 inches high was enlarged so that it is 12 inches wide. How high is the enlargement?

14. Philip has been eating 2 hamburgers every 5 days. At that rate, how many hamburgers will he eat in 30 days?

Practice

Exploring Similar Polygons

In the figure at the right, △ABC is similar to △DEF.

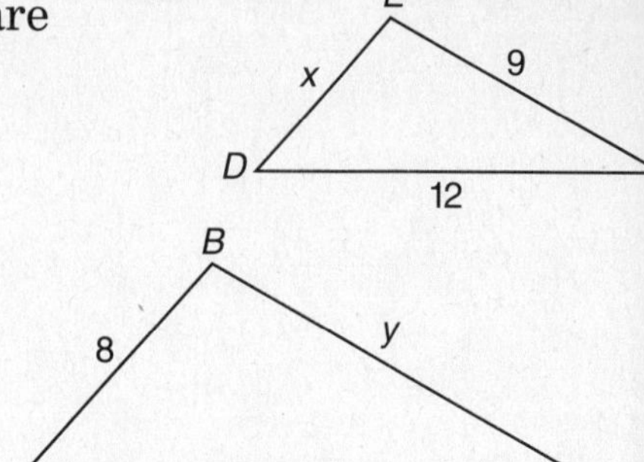

1. Write three equal ratios to show corresponding sides are proportional.

2. Find the value of x.

3. Find the value of y.

4. Find the ratio $\frac{m\angle A}{m\angle D}$.

In the figure at the right, quadrilateral ABCD is similar to quadrilateral EFGH.

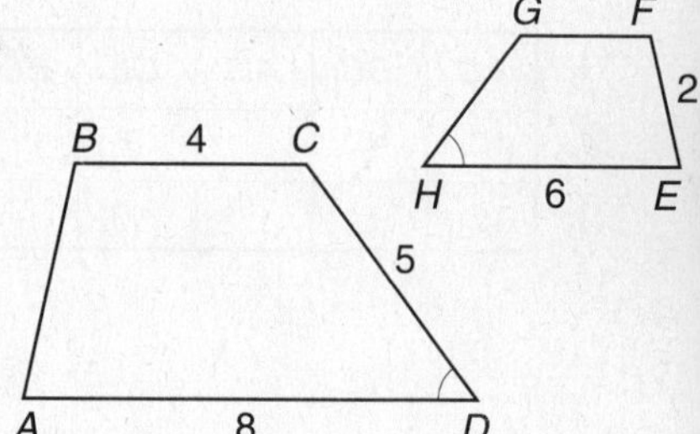

5. Write four equal ratios to show corresponding sides are proportional.

6. Find AB.

7. Find HG.

8. Find FG.

9. The sum of the measures of $\angle A$ and $\angle C$ equals the sum of the measures of which two angles of quadrilateral $EFGH$?

7–3

NAME ________________________________ DATE ____________

Practice

Student Edition
Pages 354–361

Identifying Similar Triangles

Identify the similar triangles in each figure. Explain why they are similar and use the given information to find x and y.

1.

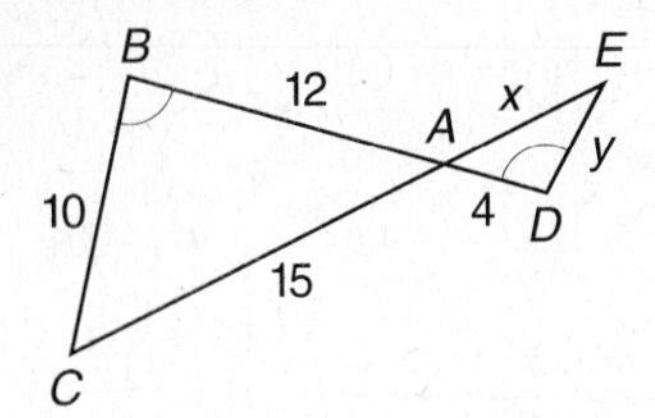

2.

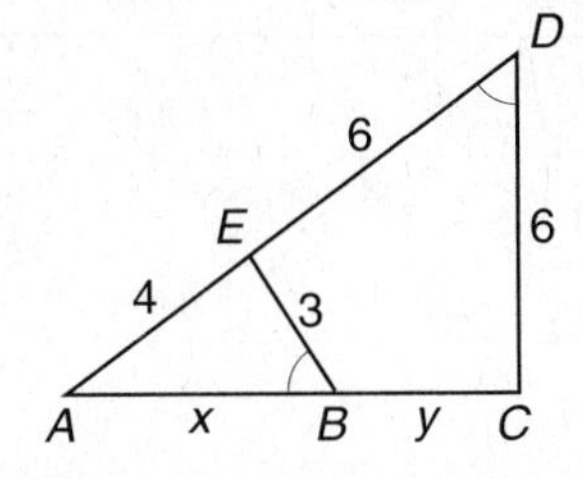

3.

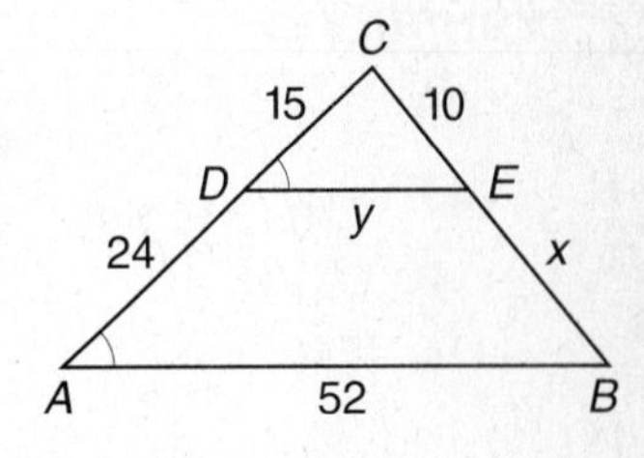

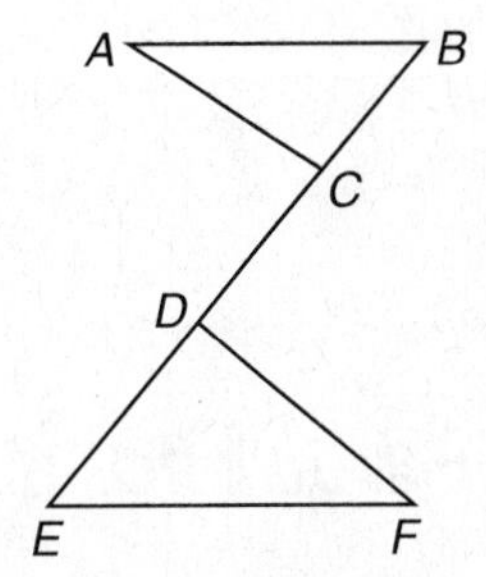

Write a two-column proof.

4. Given: $\overline{AB} \parallel \overline{EF}$
$\overline{AC} \parallel \overline{DF}$

Prove: $\triangle ABC \sim \triangle FED$

Proof:

Statements	Reasons

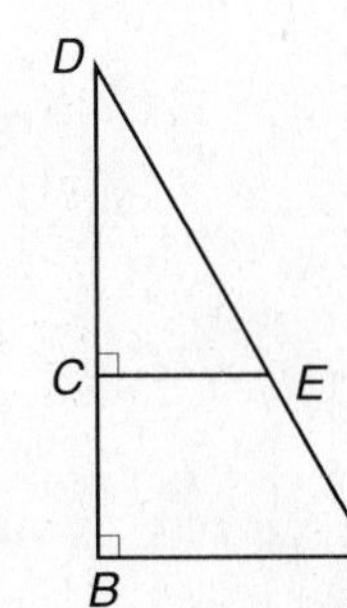

5. Given: $\overline{AB} \perp \overline{BD}$
$\overline{ED} \perp \overline{BD}$

Prove: $\triangle BDA \sim \triangle CDE$

Proof:

Statements	Reasons

Parallel Lines and Proportional Parts

Refer to the figure at the right for Exercises 1–2. Determine whether it is always true that $\overline{AB} \parallel \overline{YZ}$ under the given conditions.

1. $XA = 6$
   $AY = 4$
   $XB = 8$
   $BZ = 5$

2. $XB = 3$
   $BZ = 2$
   $AB = 6$
   $YZ = 10$

In $\triangle PQR$, find x and y so that $\overline{JG} \parallel \overline{RQ}$.

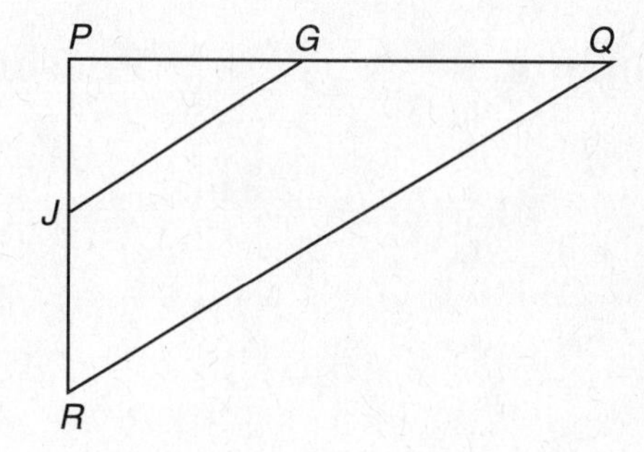

3. $PJ = 6$
   $JG = 5$
   $PG = 4$
   $GQ = x$
   $RQ = x + 6$
   $JR = y$

4. $RQ = 10$
   $JG = 8$
   $PJ = 8x - 5$
   $JR = x$
   $PG = 3y + 2$
   $QG = y$

5. In the figure at the right, $\overleftrightarrow{YA} \parallel \overleftrightarrow{OE} \parallel \overleftrightarrow{BR}$. Find the values of x and y if $YO = 4$, $ER = 16$, and $AR = 24$.

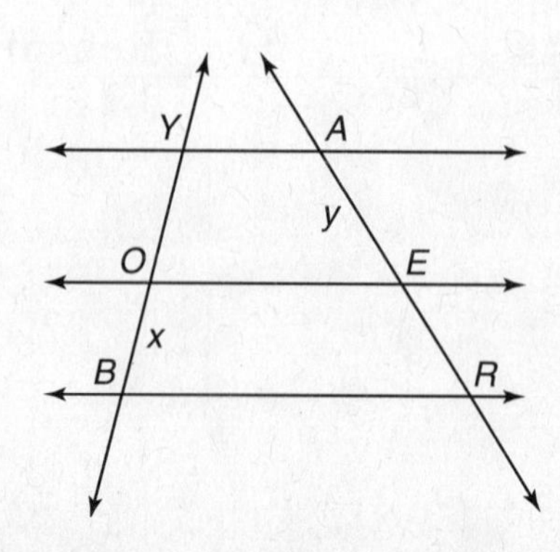

NAME ______________________ DATE ____________

Practice

Parts of Similar Triangles

In the figure at the right, $\triangle ABC \sim \triangle DEF$, $\overline{BR} \cong \overline{RC}$, and $\overline{ES} \cong \overline{SF}$. Find the value of x.

1. $BC = 24$
$EF = 15$
$AR = x$
$DS = x - 6$

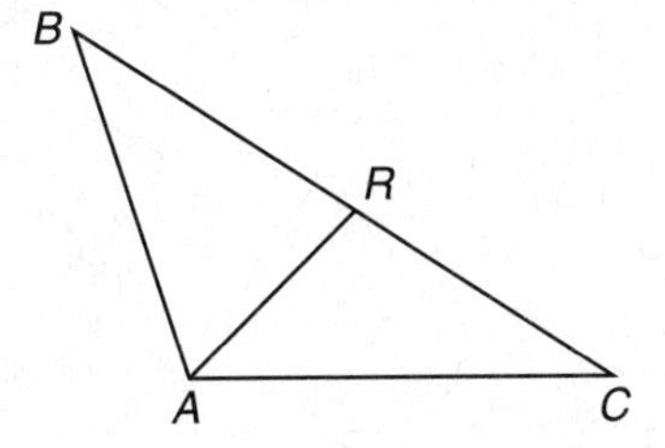

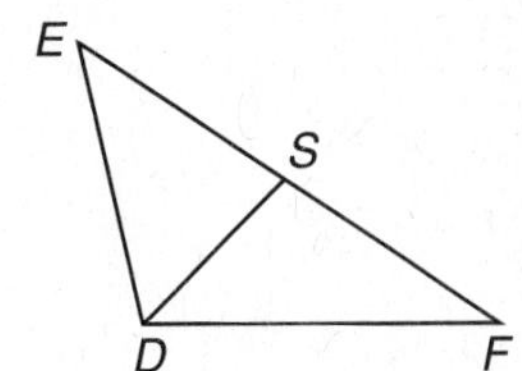

2. $AB = 2x + 5$
$DE = x + 7$
$AR = 24$
$DS = 18$

Find the value of x.

3.

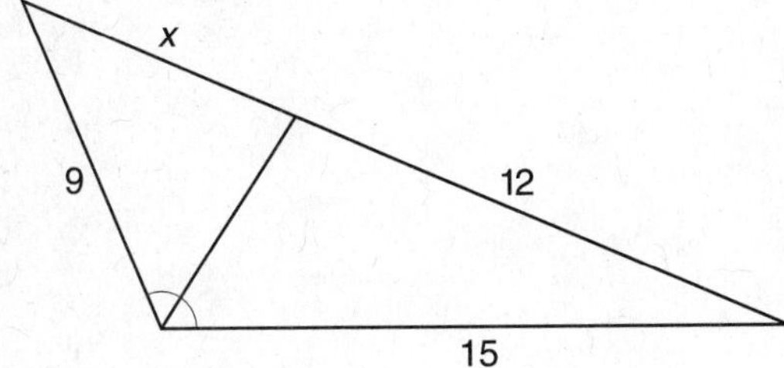

4.

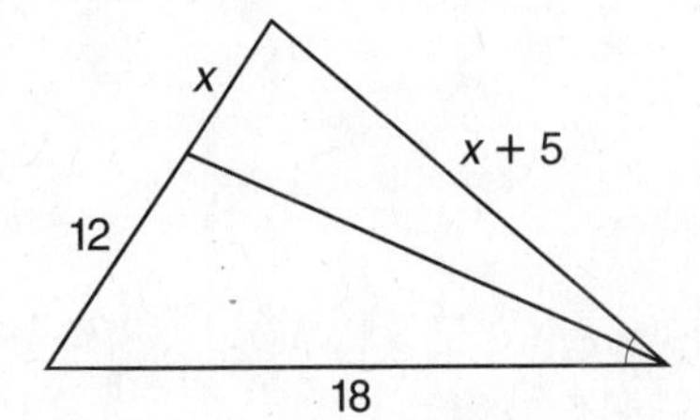

In the figure at the right, $\triangle ABC \sim \triangle DEF$, and $\overline{BX}$ and $\overline{EY}$ are altitudes. Find the value of x.

5. $AB = 25$
$DE = 16$
$BX = 18$
$EY = x$

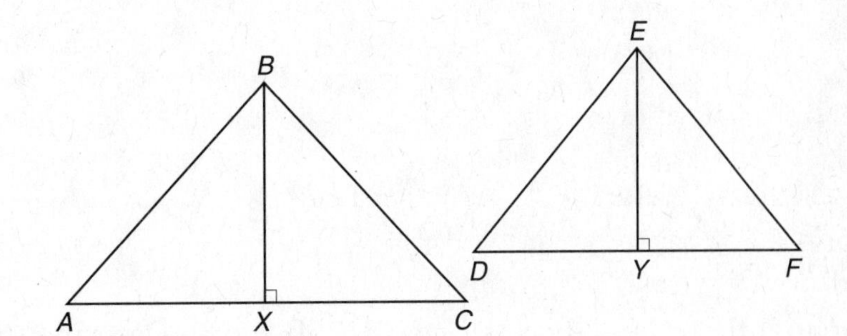

6. $AB = 30$
$DE = 25$
$BX = 2x + 5$
$EY = x + 10$

NAME ________________________________ DATE ____________

Practice

Fractals and Self-Similarity

Use the drawings below of three stages of a variation on the Koch curve to complete Exercises 1 and 2.

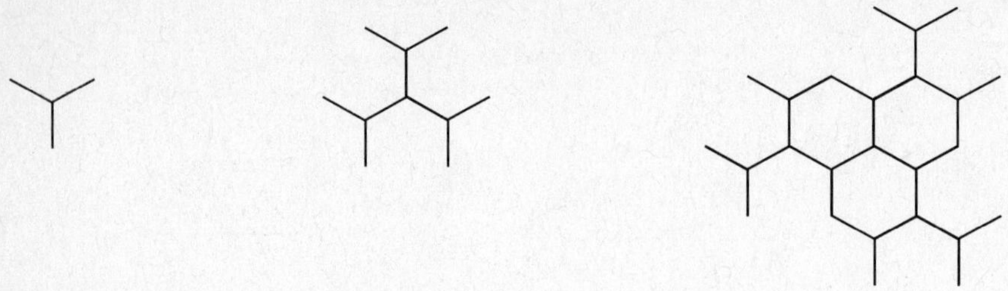

Stage 1 Stage 2 Stage 3

1. Draw Stage 4.

2. Describe the iterative process used in this variation.

3. Draw an equilateral triangle. Divide each side into fourths, and connect the points to form three lines parallel to each side of the triangle.

4. Find the number of similar triangles within the figure in Exercise 3.

5. Create a figure and draw stages 1–3 of iteration.

6. Describe the iterative process used in Exercise 5.

7. Solve a Simpler Problem How many diagonals can be drawn for a polygon with 15 sides?

NAME________________________________ DATE ____________

Practice

Student Edition
Pages 397–404

Geometric Mean and the Pythagorean Theorem

Find the geometric mean between each pair of numbers.

1. 5 and 10

2. 3 and 27

3. 6 and $\frac{1}{2}$

4. 16 and $\frac{1}{9}$

Find the values of x and y.

5.

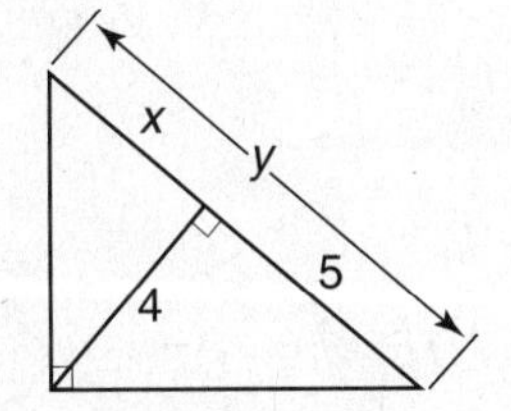

6.

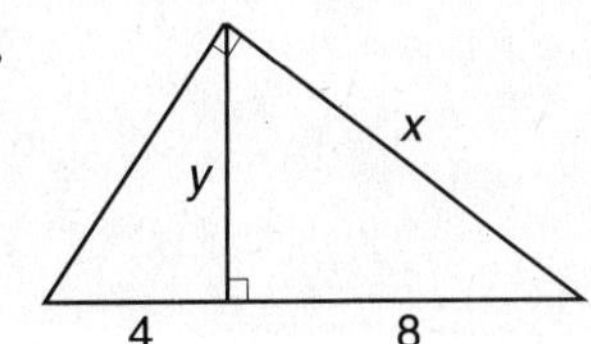

7.

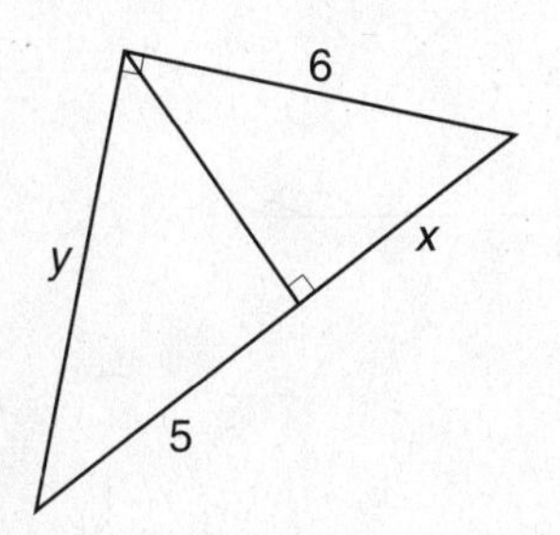

8.

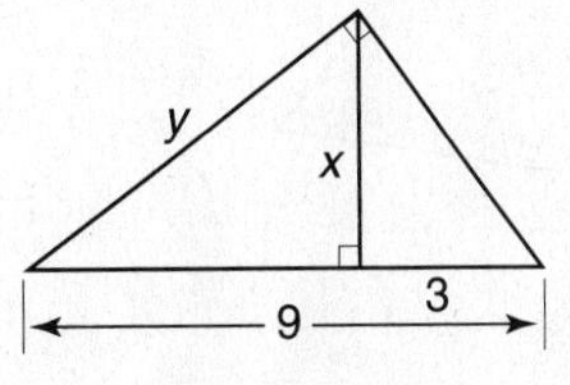

9.

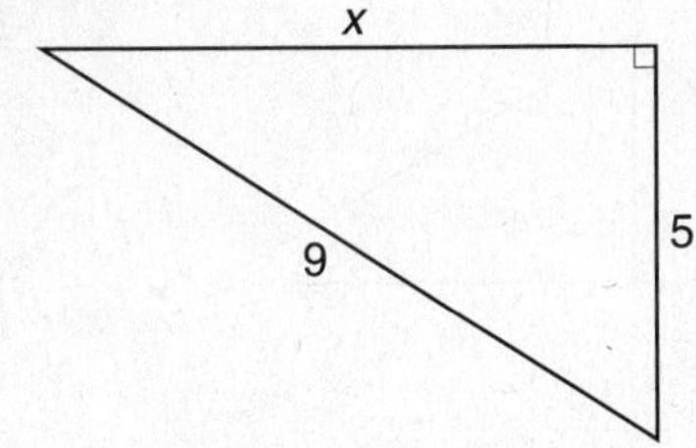

10.

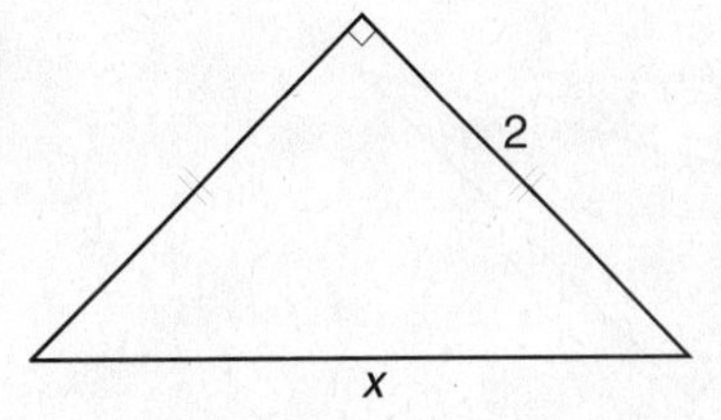

11.

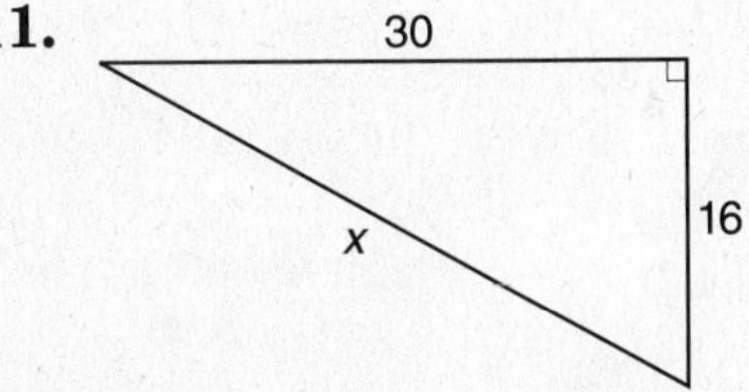

12.

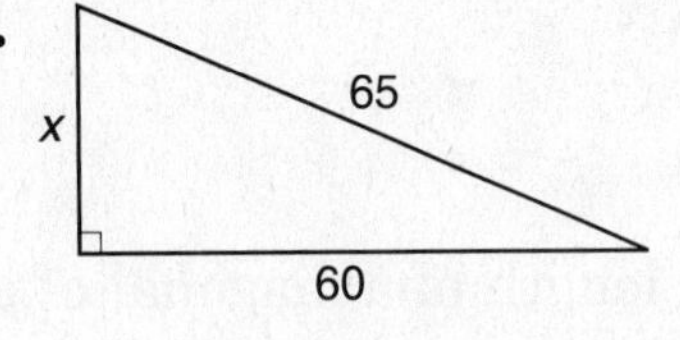

Determine if the given measures are measures of the sides of a right triangle.

13. 14, 48, 50

14. 50, 75, 85

15. 15, 36, 39

16. 45, 60, 80

Practice

Special Right Triangles

Find the values of x and y.

1.

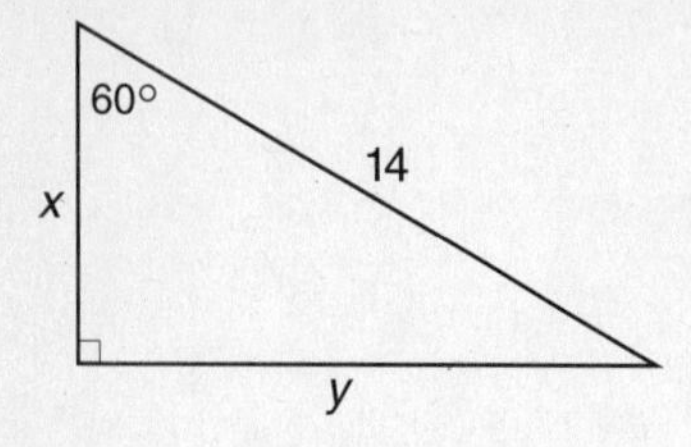

2.

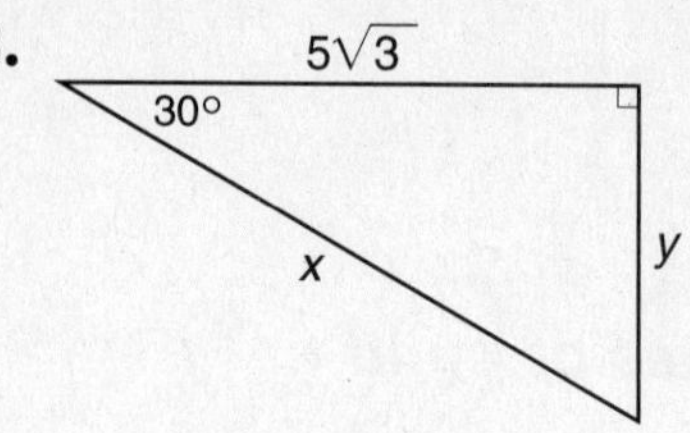

3.

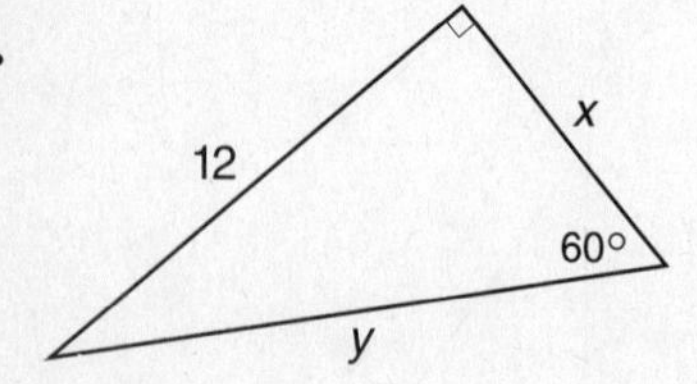

4.

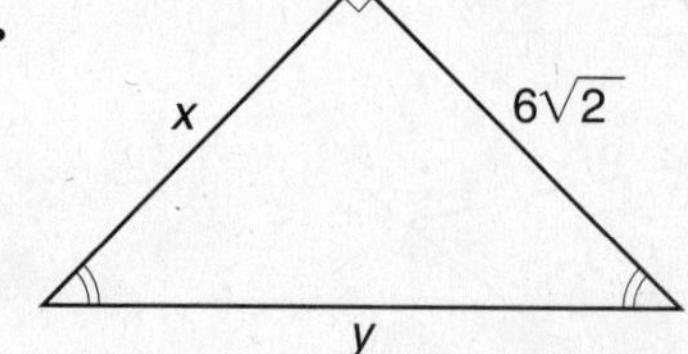

5.

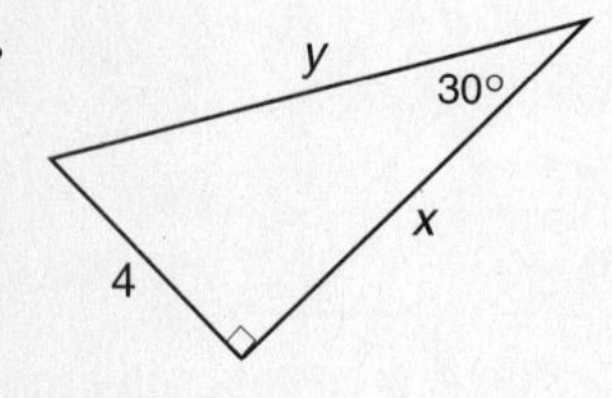

6.

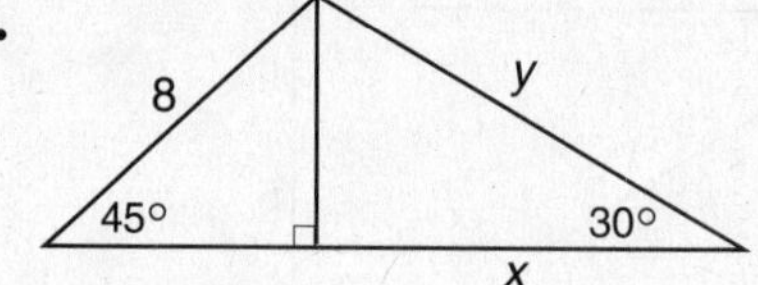

7. Find the length of a diagonal of a square with sides 10 in. long.

8. Find the length of a side of a square whose diagonal is 4 cm.

9. One side of an equilateral triangle measures 6 cm. Find the measure of an altitude of the triangle.

NAME ______________________ DATE ____________

Practice

Integration: Trigonometry Ratios in Right Triangles

Find the values of x and y. Round to the nearest tenth.

1.

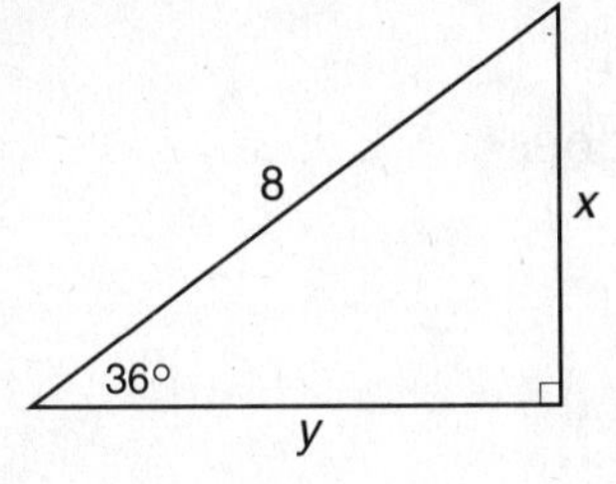

2.

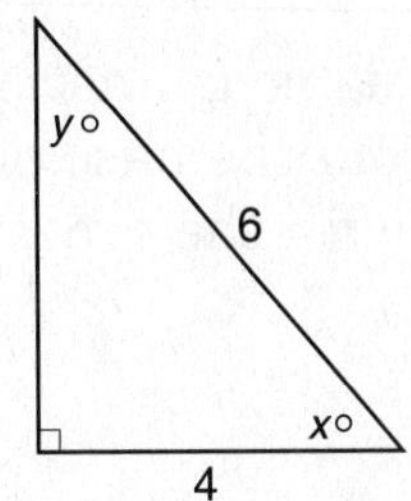

3.

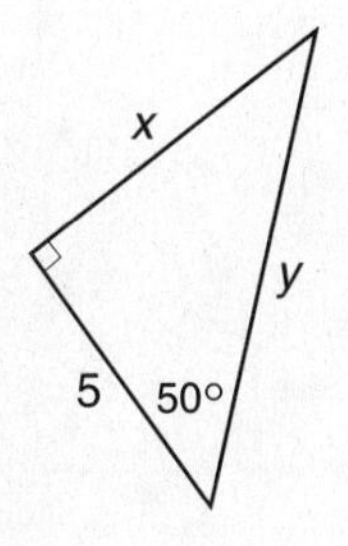

4.

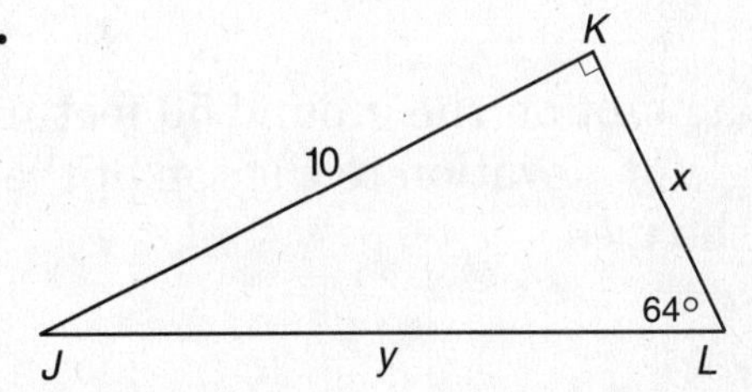

5.

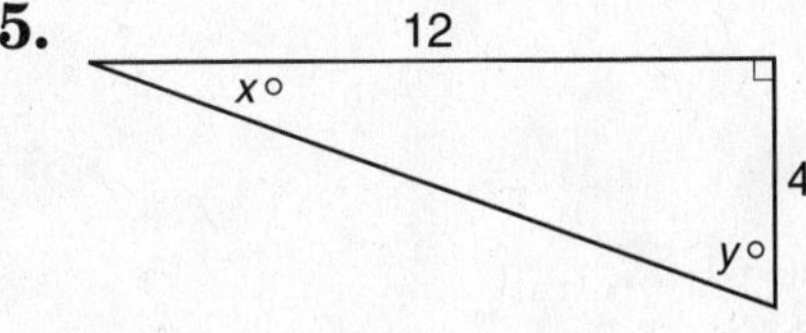

6.

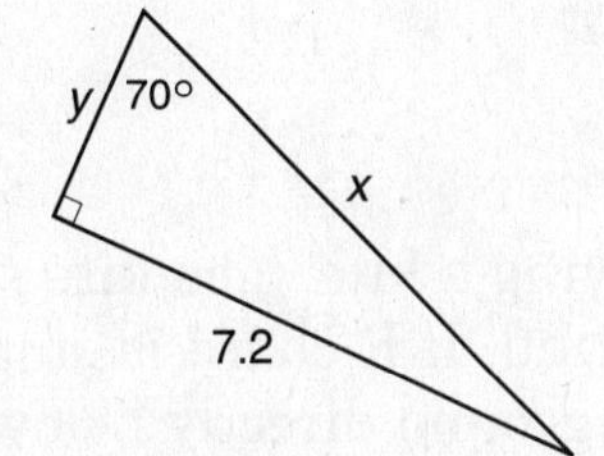

Practice

Angles of Elevation and Depression

Solve each problem. Round measures of segments to the nearest hundredth and measures of angles to the nearest degree.

1. A 20-foot ladder leans against a wall so that the base of the ladder is 8 feet from the base of the building. What angle does the ladder make with the ground?

2. A 50-meter vertical tower is braced with a cable secured at the top of the tower and tied 30 meters from the base. What angles does the cable form with the vertical tower?

3. At a point on the ground 50 feet from the foot of a tree, the angle of elevation to the top of the tree is 53°. Find the height of the tree.

4. From the top of a lighthouse 210 feet high, the angle of depression to a boat is 27°. Find the distance from the boat to the foot of the lighthouse. The lighthouse was built at sea level.

5. Richard is flying a kite. The kite string makes an angle of 57° with the ground. If Richard is standing 100 feet from the point on the ground directly below the kite, find the length of the kite string.

6. An airplane rises vertically 1000 feet over a horizontal distance of 1 mile. What is the angle of elevation of the airplane's path? (Hint: 1 mile = 5280 feet)

Using the Law of Sines

Solve each $\triangle ABC$. Round measures to the nearest tenth.

1. $a = 12, m\angle B = 70, m\angle C = 15$

2. $a = 12, b = 5, m\angle A = 110$

3. $a = 8, m\angle A = 60, m\angle C = 40$

4. $a = 5, c = 4, m\angle A = 65$

5. $b = 6, m\angle A = 44, m\angle B = 68$

6. $a = 7, m\angle A = 37, m\angle B = 76$

7. $a = 9, b = 9, m\angle C = 20°$

8. A ship is sighted from two radar stations 43 km apart. The angle between the line segment joining the two stations and the radar beam of the first station is 37°. The angle between the line segment joining the two stations and the beam from the second station is 113°. How far is the ship from the second station?

Practice

Using the Law of Cosines

Solve each triangle △ABC described below. Round measures to the nearest tenth.

1. $a = 16, b = 20, m\angle B = 40$

2. $a = 10, b = 15, c = 12$

3. $a = 42, c = 60, m\angle B = 58$

4. $m\angle A = 60, m\angle B = 72, c = 9$

5. $a = 7, b = 12, c = 15$

6. $m\angle A = 43, b = 23, c = 26$

7. $a = 16, m\angle A = 23, m\angle B = 87$

8. $c = 15.6, a = 12.9, b = 18.4$

9. Decision-Making Complete the addition problem at the right. If a letter is used more than once, it represents the same digit each time.

```
  SOME
 +MORE
 -----
 SENSE
```

NAME ______________________ DATE ____________

Practice

Student Edition
Pages 446–451

Exploring Circles

Refer to the figure at the right.

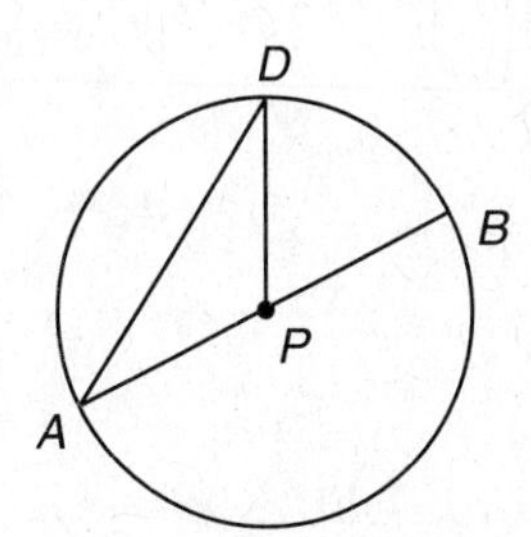

1. Name the center of $\odot P$.

2. Name the three radii of the circle.

3. Name a diameter.

4. Name two chords.

Find the circumference of a circle with a radius of the given length. Round your answers to the nearest tenth.

5. 3 cm

6. 2 ft

7. 34 mm

8. 4.5 m

Find the exact circumference of each circle.

9.

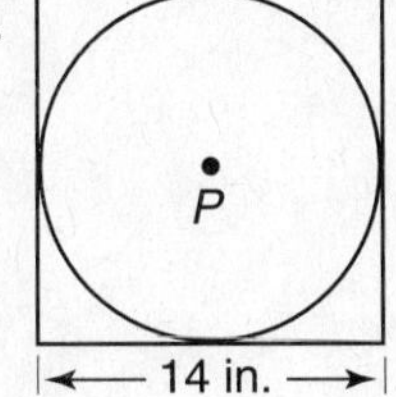

10.

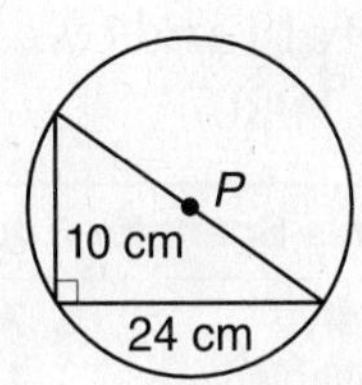

11.

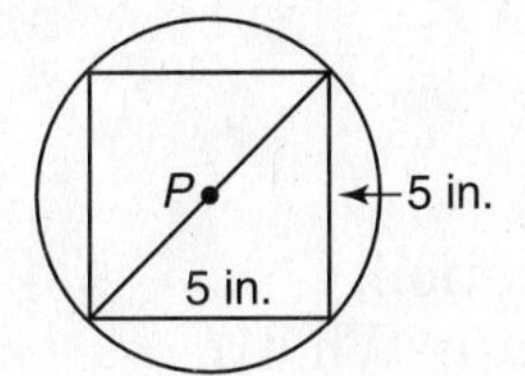

12.

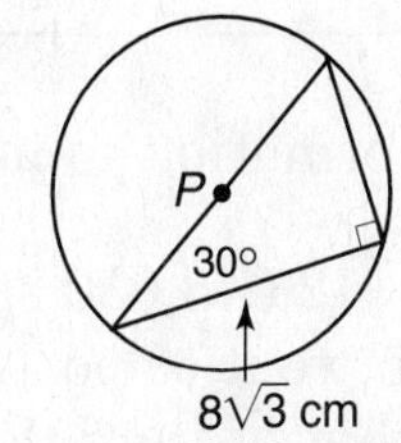

Practice

Angles and Arcs

In ⊙P, m∠1 = 140 with diameter $\overline{AC}$. Find each measure.

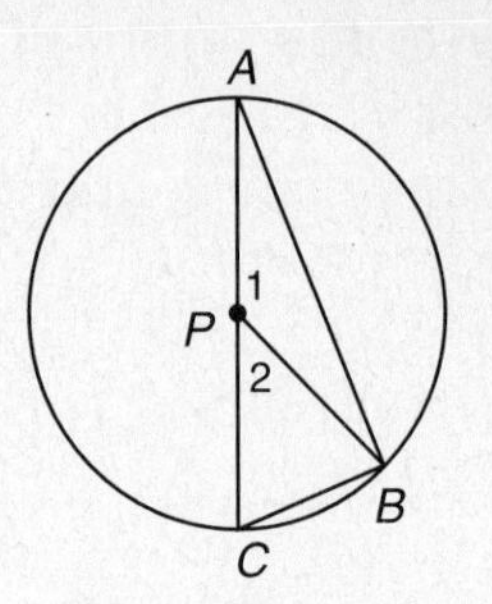

1. $m\angle 2$
2. $m\widehat{BC}$
3. $m\widehat{AB}$
4. $m\widehat{ABC}$

In ⊙P, m∠2 = m∠1, m∠2 = 4x + 35, m∠1 = 9x + 5 with diameters $\overline{BD}$ and $\overline{AC}$. Find each of the following.

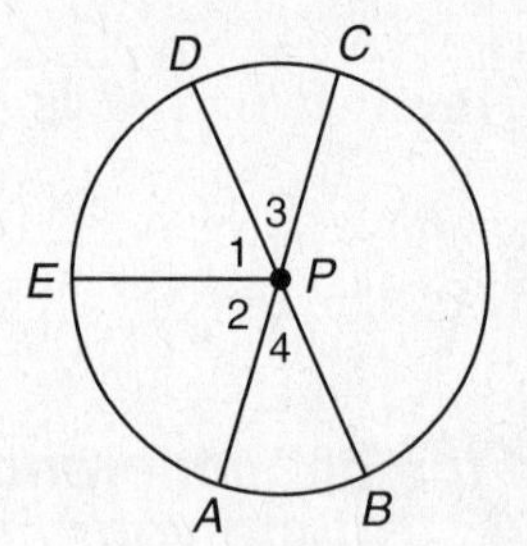

5. x
6. $m\widehat{AE}$
7. $m\widehat{ED}$
8. $m\angle 3$
9. $m\widehat{AB}$
10. $m\widehat{EC}$
11. $m\widehat{EB}$
12. $m\angle CPB$
13. $m\widehat{CB}$
14. $m\widehat{CEB}$
15. $m\widehat{DC}$
16. $m\widehat{CEA}$

17. In ⊙A, $AB = 12$ and $m\angle BAC = 60$. Find the length of $\widehat{BC}$.

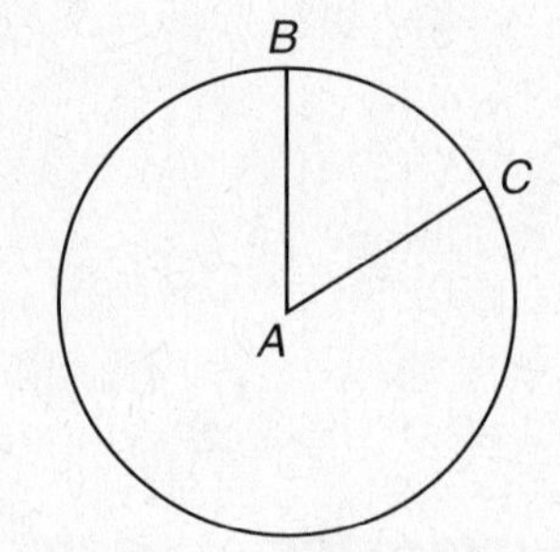

18. **Using Graphs** The table below shows how federal funds were spent on education in 1990.

1990 Federal Funds Spent for Education	
Elementary/Secondary	$ 7,945,177
Education for the Disabled	4,204,099
Post-Secondary Education	12,645,630
Public Library Services	145,367
Other	760,616
Total	$25,700,889

a. Use the information to make a circle graph.

b. Out of the $12,645,630 spent on post-secondary education, $10,801,185 went to post-secondary financial assistance. What percent is that of the $12,645,630?

NAME______________________________ DATE ____________

Practice

Student Edition
Pages 459–465

Arcs and Chords

In each figure, O is the center. Find each measure to the nearest tenth.

1. $m\widehat{BC}$

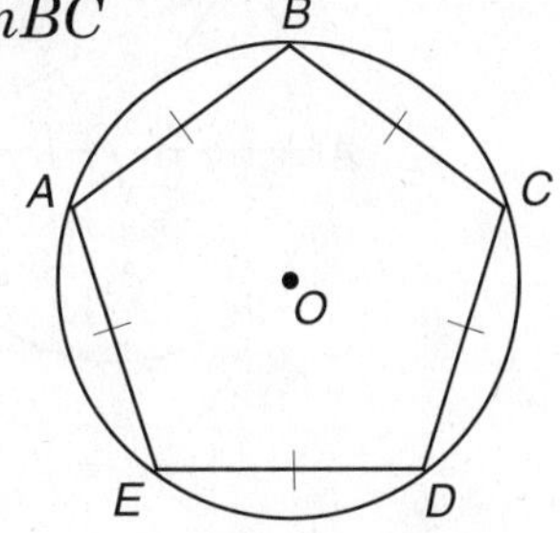

2. YQ

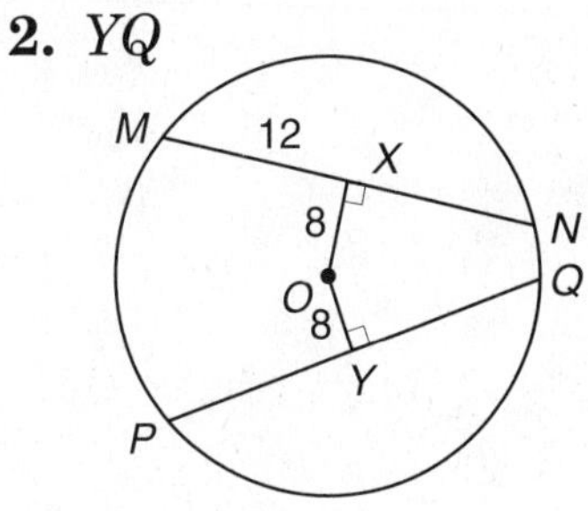

3. Suppose a chord of a circle is 16 inches long and is 6 inches from the center of the circle. Find the length of a radius.

4. Find the length of a chord that is 5 inches from the center of a circle with a radius of 13 inches.

5. Suppose a radius of a circle is 17 units and a chord is 30 units long. Find the distance from the center of the circle to the chord.

6. Find AB.

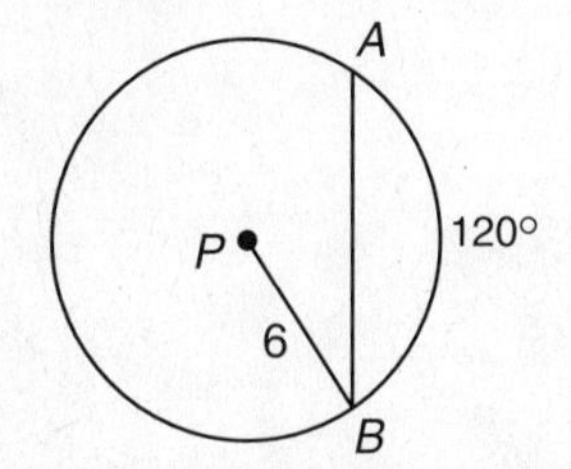

7. Find AB

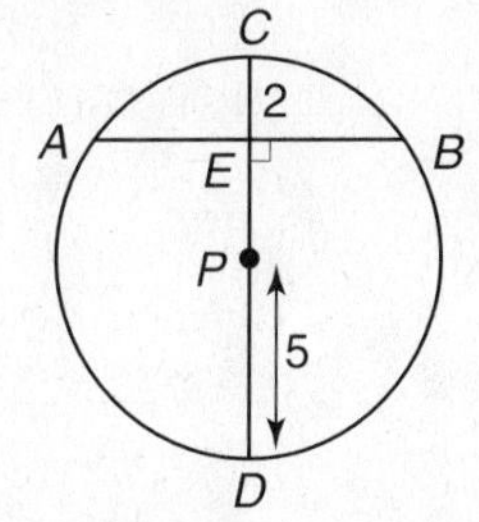

Inscribed Angles

In $\odot P$, $m\widehat{AB} = x$ and $m\widehat{BC} = 3x$. Find each measure.

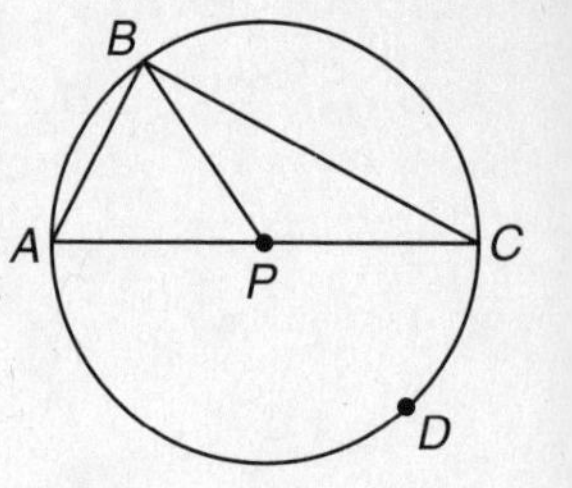

1. $m\widehat{ADC}$

2. $m\widehat{AB}$

3. $m\widehat{BC}$

4. $m\angle ABC$

5. $m\angle A$

6. $m\angle C$

In $\odot Q$, $m\angle ABC = 72$ and $m\widehat{CD} = 46$. Find each measure.

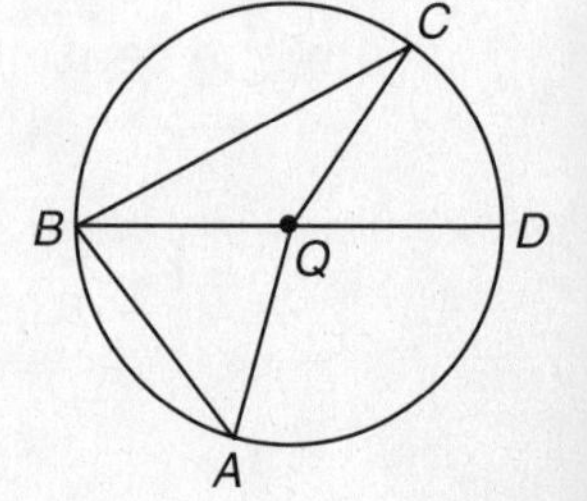

7. $m\widehat{CA}$

8. $m\widehat{AD}$

9. $m\angle ABD$

10. $m\widehat{BC}$

11. $m\angle C$

12. $m\angle A$

13. Suppose $ABCD$ is a trapezoid that has its vertices on $\odot P$, with $AB \parallel CD$. Write a paragraph proof to show that $ABCD$ is an isosceles trapezoid.

NAME ______________________________ DATE ____________

Practice

Tangents

For each ⊙Q, find the value of x. Assume that segments that appear to be tangent are tangent.

1.

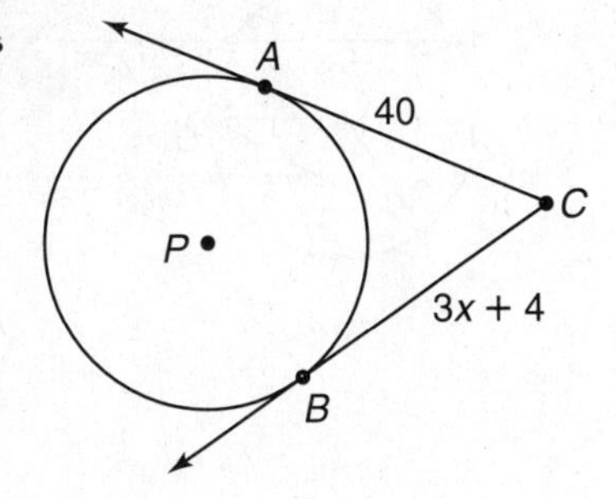

2.

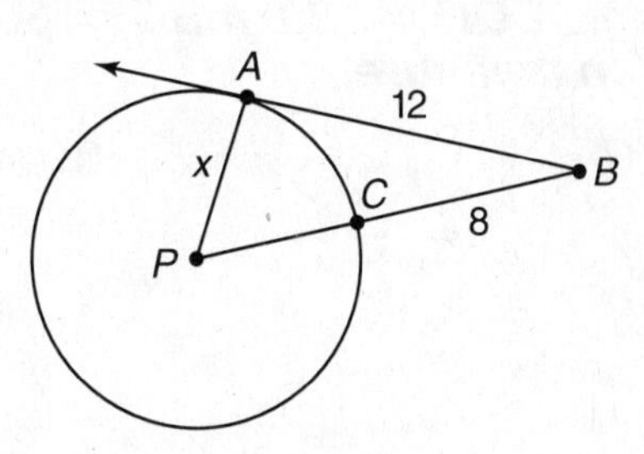

3.

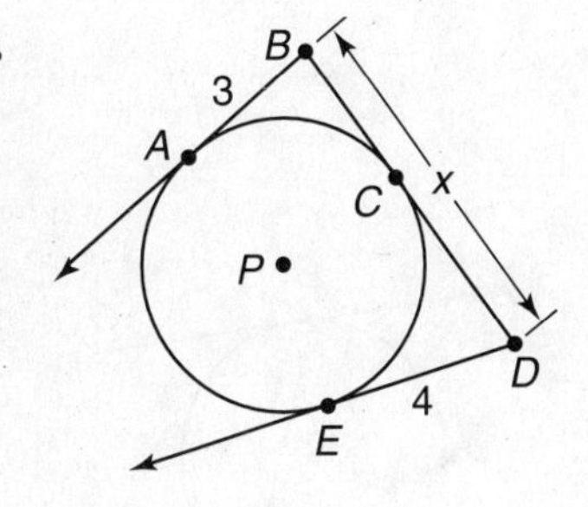

4.

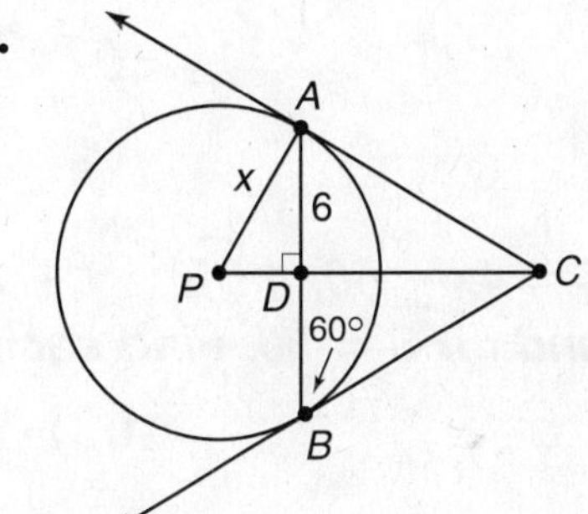

5.

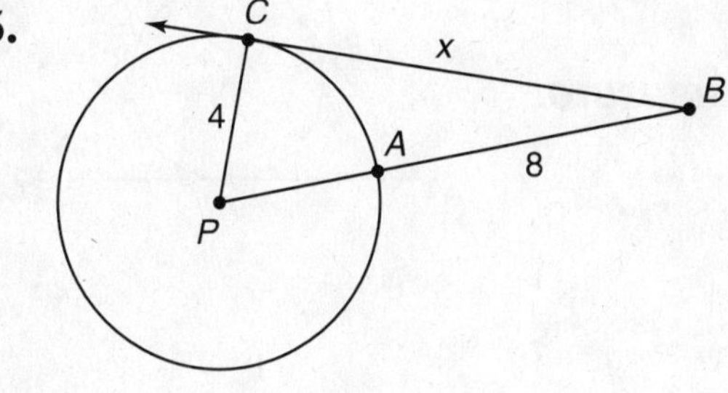

6.

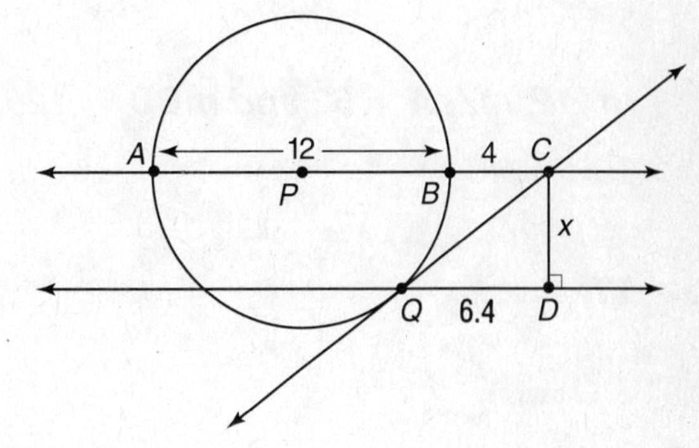

NAME ______________________________ DATE ____________

Practice

Secants, Tangents, and Angle Measures

Assume that lines that appear to be tangents are tangents.

In ⊙Q, $m\angle CQD = 120$, $m\widehat{BC} = 30$, and $m\angle BEC = 25$. Find each measure.

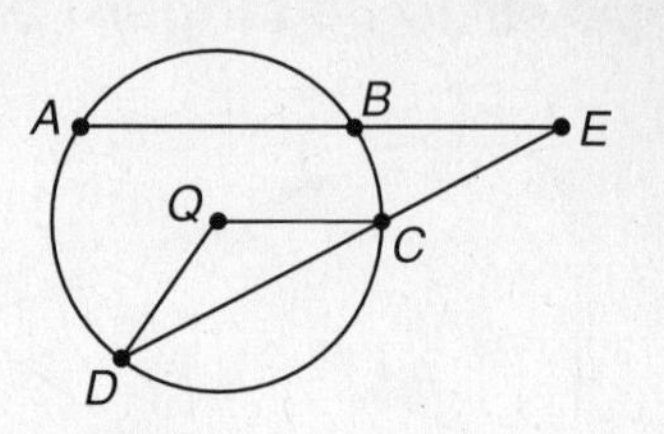

1. $m\widehat{DC}$

2. $m\widehat{AD}$

3. $m\widehat{AB}$

4. $m\angle QDC$

In ⊙Q, $m\widehat{AE} = 140$, $m\widehat{BD} = y$, $m\widehat{AB} = 2y$, and $m\widehat{DE} = 2y$. Find each measure.

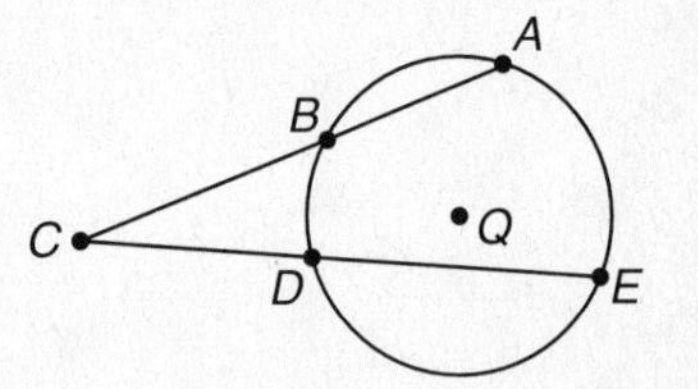

5. $m\widehat{BD}$

6. $m\widehat{AB}$

7. $m\widehat{DE}$

8. $m\angle BCD$

In ⊙P, $m\widehat{BC} = 4x - 50$, $m\widehat{DE} = x + 25$, $m\widehat{EF} = x - 15$, $m\widehat{CD} = x$, and $m\widehat{FB} = 50$. Find each measure.

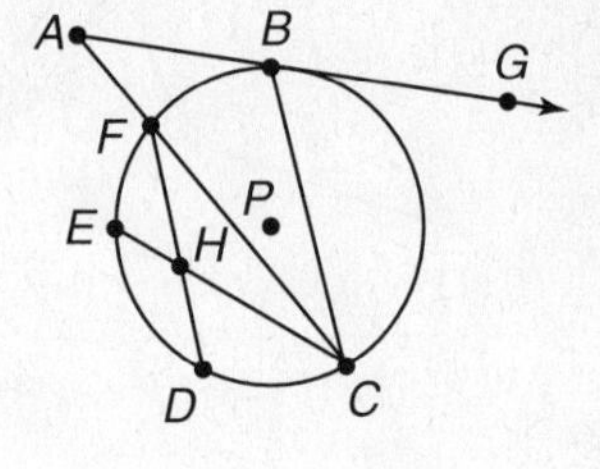

9. $m\angle A$

10. $m\angle BCA$

11. $m\angle ABC$

12. $m\angle GBC$

13. $m\angle FHE$

14. $m\angle CFD$

In ⊙P, $m\angle A = 62$ and $m\widehat{BD} = 120$. Find each measure.

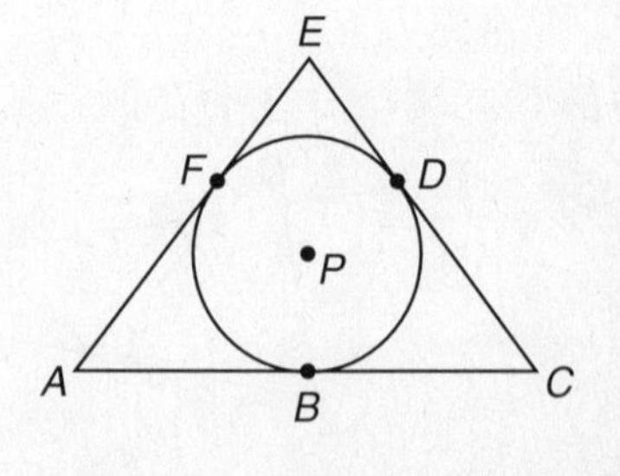

15. $m\angle C$

16. $m\widehat{DF}$

17. $m\angle E$

Practice

Special Segments in a Circle

Find the value of x to the nearest tenth. Assume segments that appear to be tangents are tangents.

1.

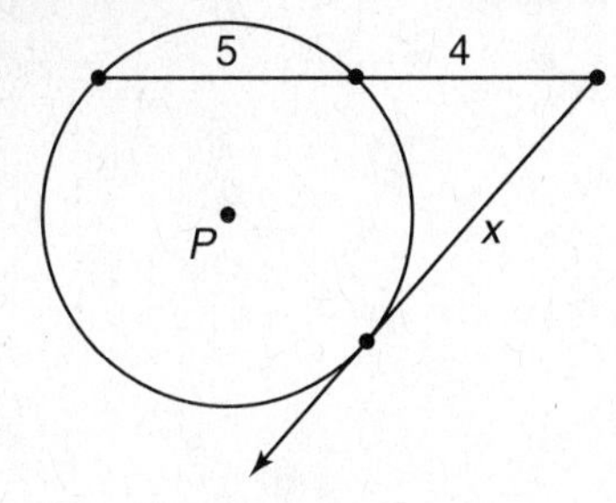

2.

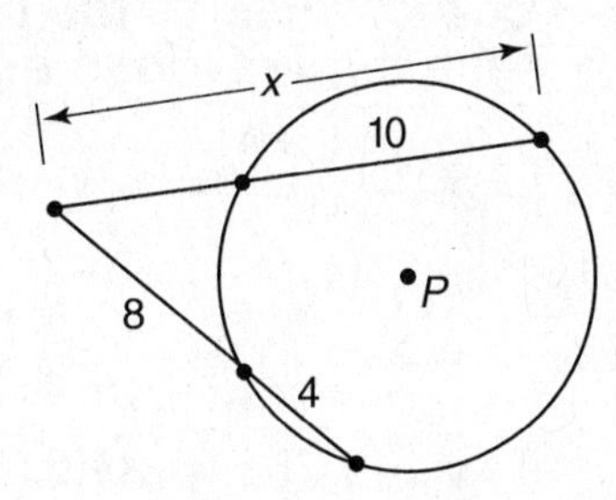

3.

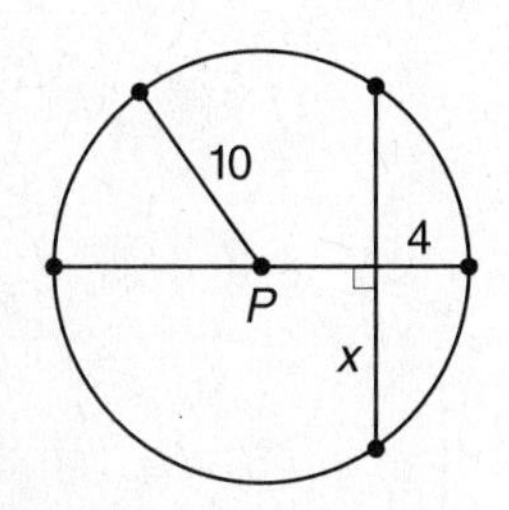

4.

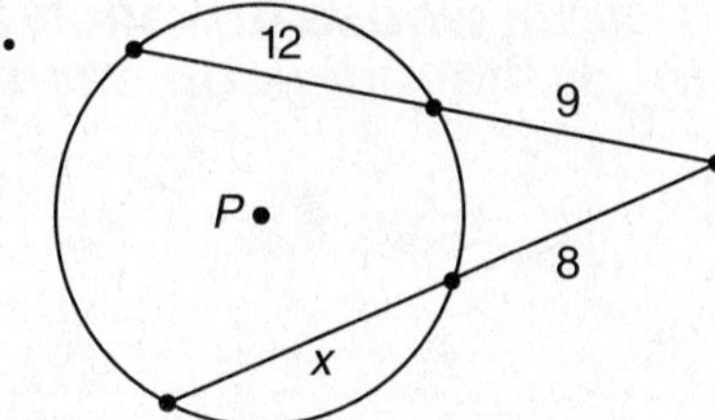

In ⊙P, CE = 6, CD = 16, and AB = 17. Find each measure.

5. EB

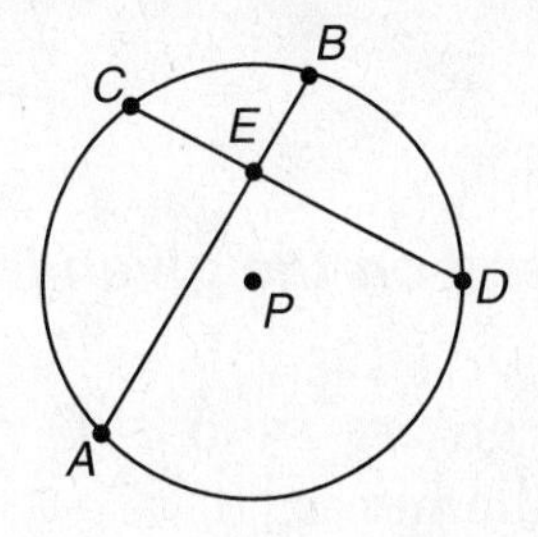

6. AE

In ⊙P, AC = 3, BC = 5, and AD = 2. Find each measure.

7. PD

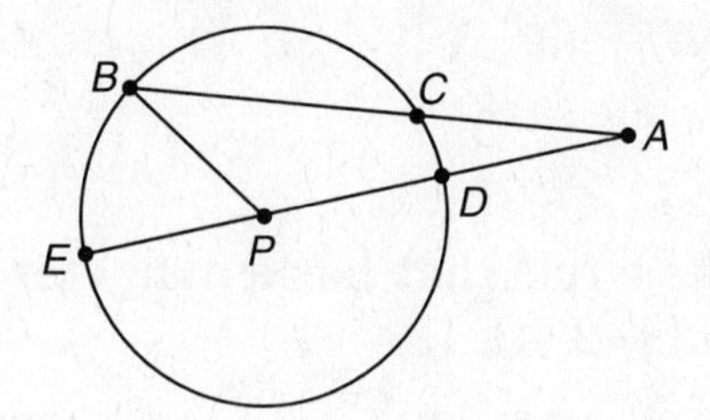

8. ED

9. PB

NAME________________________________ DATE ____________

Practice

Integration: Algebra Equations of Circles

Determine the coordinates of the center and the measure of the radius for each circle whose equation is given.

1. $(x - 3)^2 + (y + 1)^2 = 16$

2. $\left(x + \frac{5}{8}\right)^2 + (y + 2)^2 - \frac{25}{9} = 0$

3. $(x - 3.2)^2 + (y - 0.75)^2 = 37.21$

Graph each circle whose equation is given. Label the center and measure of the radius on each graph.

4. $(x - 2)^2 + y^2 = 6.25$

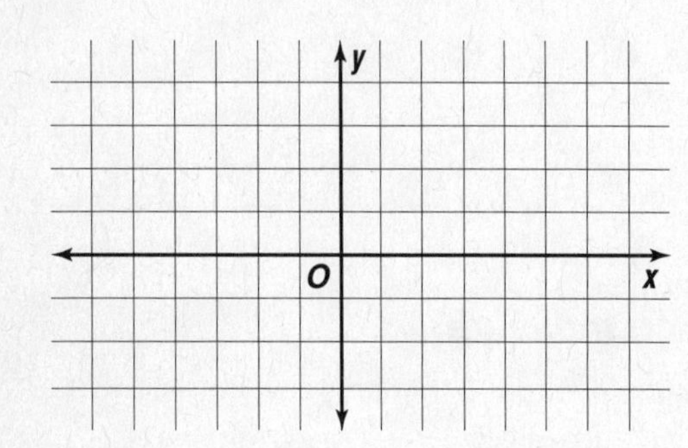

5. $(x + 3)^2 + \left(y - \frac{3}{2}\right)^2 = 4$

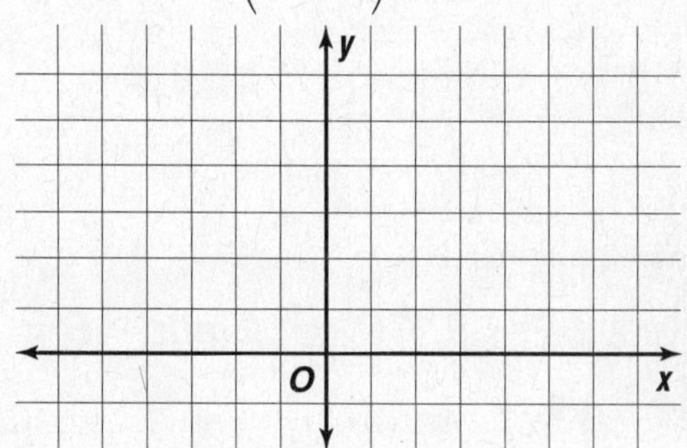

Write the equation of circle P based on the given information.

7. center: $P\left(0, \frac{1}{2}\right)$
radius: 8

8. center: $P(-5.3, 1)$
diameter: 9

9.

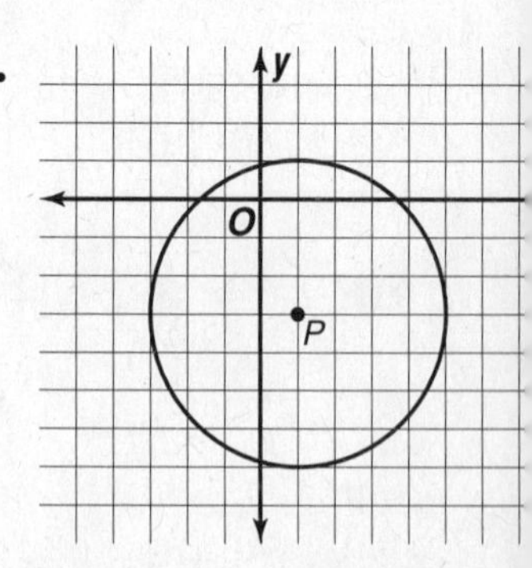

10. Write the equation of the circle that has a diameter whose endpoints are (5, −7) and (−2, 4).

Practice

Student Edition
Pages 514–521

Polygons

State the number of sides for each convex polygon.

1. quadrilateral

2. octagon

3. 83-gon

4. heptagon

5. decagon

6. hexagon

Use polygon ABCDEFG to answer each question.

7. Name the vertices of the polygon.

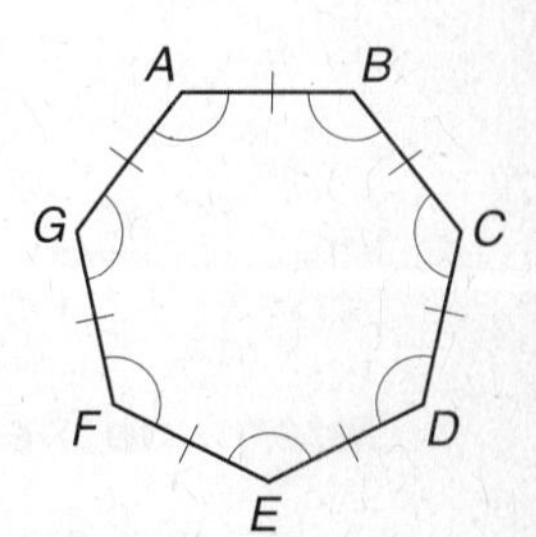

8. Name the angles of the polygon.

9. Name the sides of the polygon.

10. Is the polygon convex or concave.

11. Name the polygon according to the number of sides it has.

12. Is the polygon regular? Explain.

Find the sum of the measures of the interior angles of each convex polygon.

13. heptagon

14. octagon

15. 13-gon

The number of sides of a regular polygon is given. Find the measure of an interior and an exterior angle of the polygon.

16. 5

17. 9

18. 10

Practice

Tessellations

Determine whether each figure tessellates in a plane. If so, draw a sample figure.

1. scalene triangle

2. regular 18-gon

3. parallelogram

Determine if each pattern will tessellate.

4. octagon and isosceles triangle

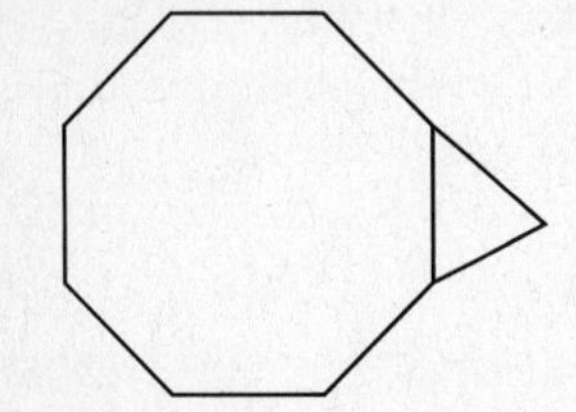

5. three isosceles triangles

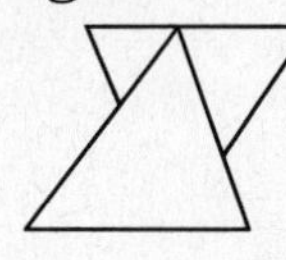

6. isosceles trapezoid anc square

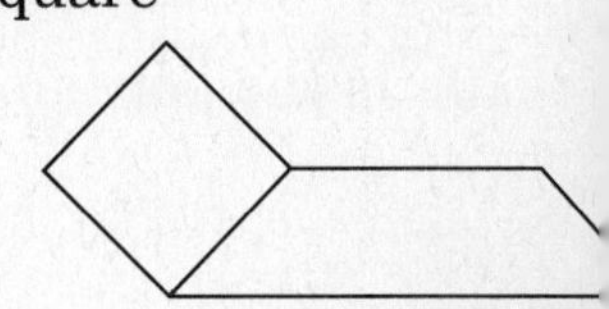

Determine whether each tessellation is regular, uniform, or semi-regular. Name all possibilities.

7. hexagon and triangle

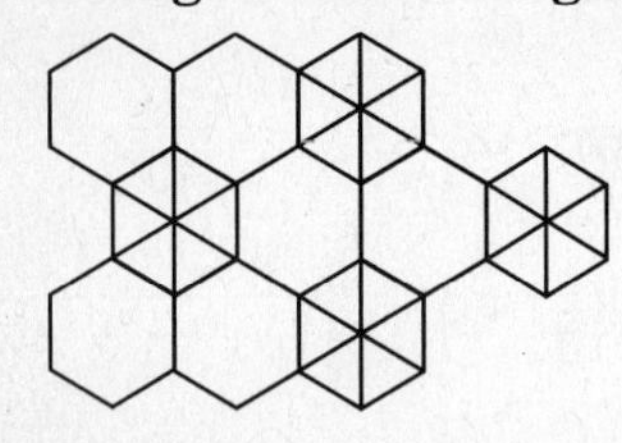

8. hexagon and triangle

9. obtuse triangle

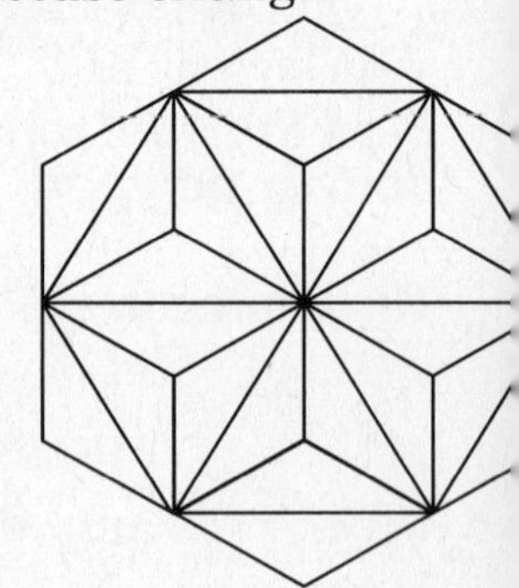

10. Guess and check Use the following three shapes and determine if they can tessellate a row. If so, draw a sample row.

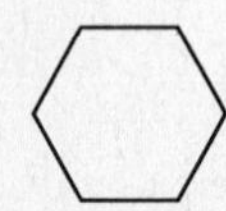
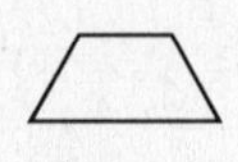

NAME ______________________________ DATE ____________

Practice

Area of Parallelograms

Find the area of each figure or shaded region. Assume that angles that appear to be right are right angles.

1.

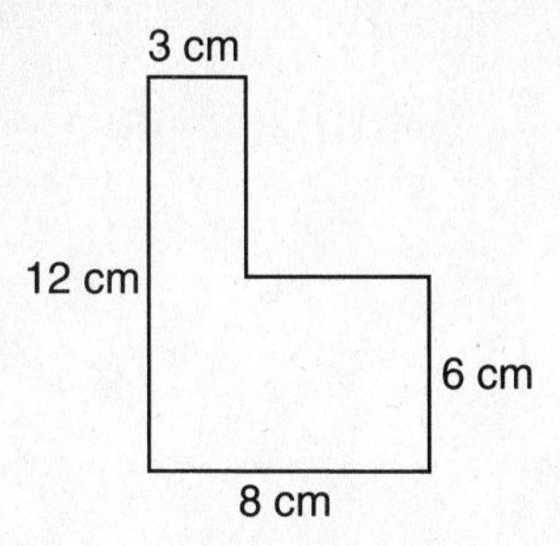

2.

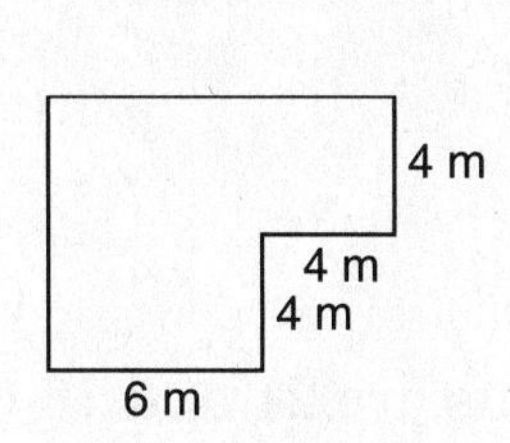

3.

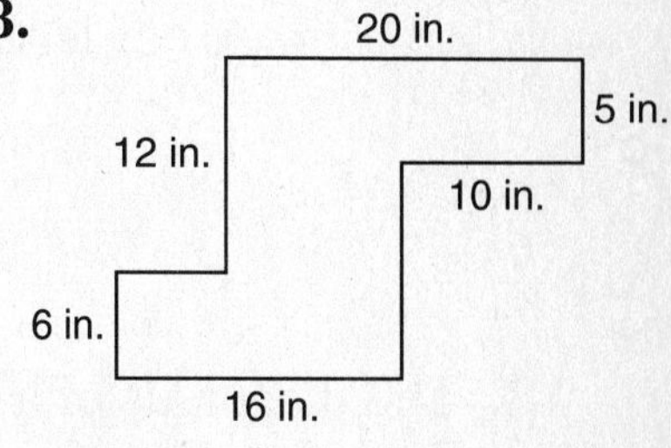

4.

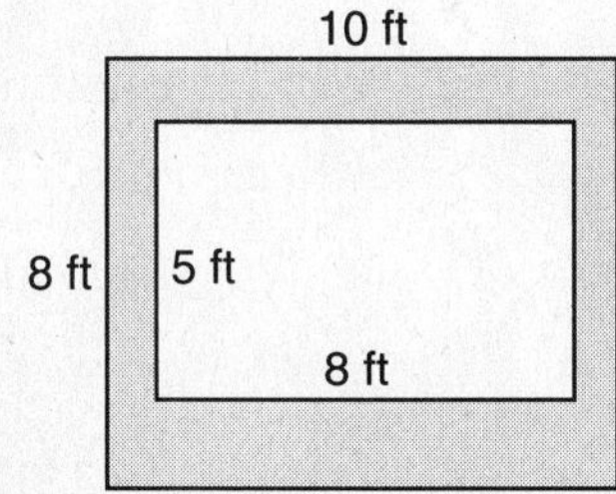

5.

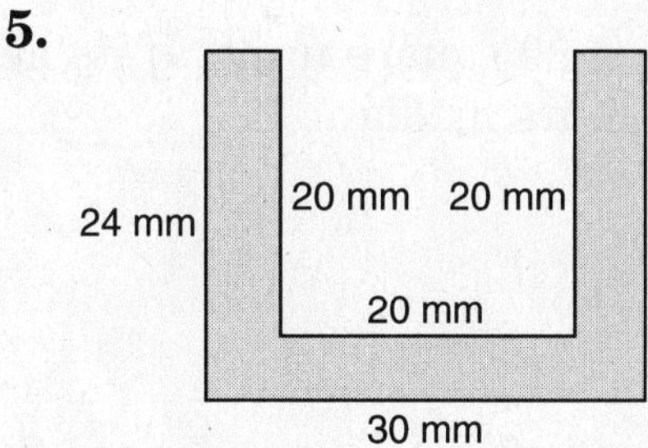

6.

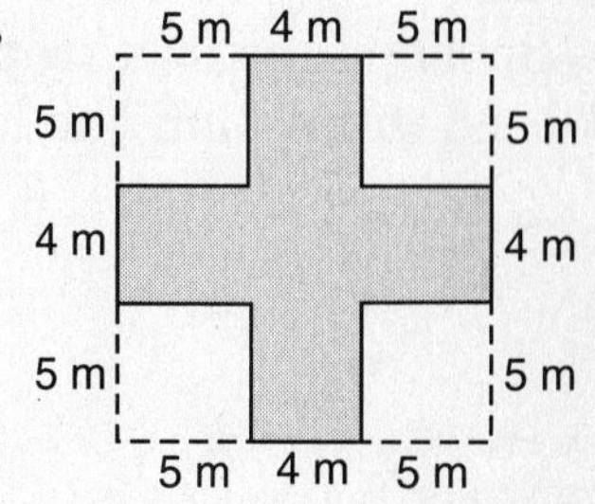

7. The sides of a parallelogram have lengths 8 inches and 16 inches and one of the angles of the parallelogram has a measure of 45°. Find the area of the parallelogram.

8. Find the area of the parallelogram that has vertices $A(0, 0)$, $B(2, 7)$, $C(10, 7)$, and $D(8, 0)$.

9. Find the area of the parallelogram that has vertices $W(-4, 15)$, $X(1, 15)$, $Y(4, 10)$, and $Z(-1, 10)$.

Area of Triangles, Rhombi, and Trapezoids

Find each missing measure.

1. The area of a triangle is 216 square units. If the height is 18 units, what is the length of the base?

2. The diagonals of a rhombus are 21 and 16 centimeters long. Find the area of the rhombus.

3. The area of a trapezoid is 80 square units. If its height is 8 units, find the length of its median.

4. The height of a trapezoid is 9 cm. The bases are 8 cm and 12 cm long. Find the area.

5. A trapezoid has an area of 908.5 cm^2. If the altitude measures 23 cm and one base measures 36 cm, find the length of the other base.

6. The measure of the consecutive sides of an isosceles trapezoid are in the ratio 8:5:2:5. The perimeter of the trapezoid is 140 inches. If its height is 28 inches, find the area of the trapezoid.

NAME______________________________ DATE ____________

Practice

Area of Regular Polygons and Circles

Find the area of each regular polygon. Round your answers to the nearest tenth.

1. an octagon with an apothem 4.8 centimeters long and a side 4 centimeters long

2. a square with a side 24 inches long and an apothem 12 inches long

3. a hexagon with a side 23.1 meters long and an apothem 20.0 meters long

4. a pentagon with an apothem 316.6 millimeters long and a side 460 millimeters long

Find the apothem, area, and perimeter of each regular polygon. Round your answers to the nearest tenth.

5.

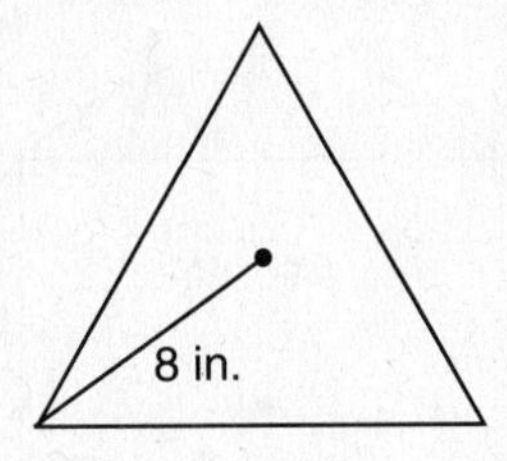

6.

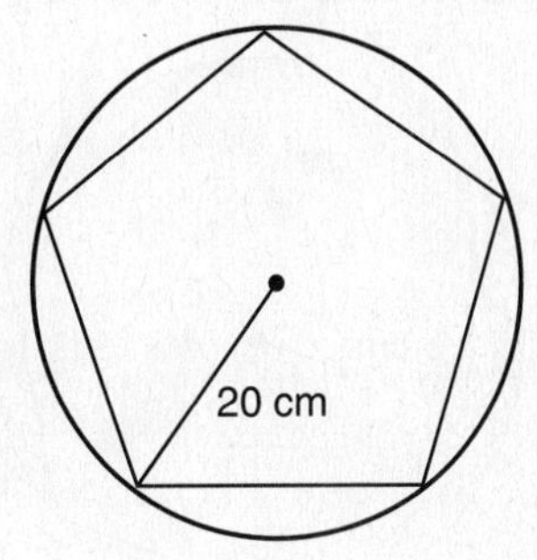

7.

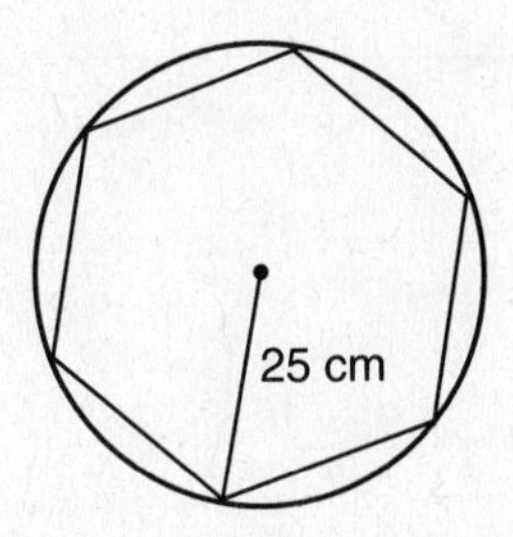

8.

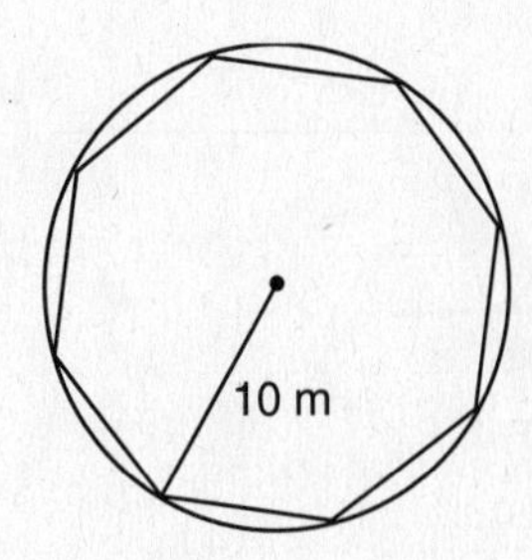

NAME ______________________________ DATE ____________

Practice

Integration: Probability: Geometric Probability

A point is chosen at random on $\overline{AB}$. C is the midpoint of $\overline{AB}$, D is the midpoint of $\overline{AC}$, and E is the midpoint of $\overline{AD}$. Find each probability.

1. The point is on $\overline{AC}$.

2. The point is on $\overline{CB}$.

3. The point is on $\overline{AD}$.

4. The point is on $\overline{DB}$.

5. The point is on $\overline{ED}$.

6. The point is on $\overline{EC}$.

Find the probability that a point chosen at random in each figure lies in the shaded region.

7.

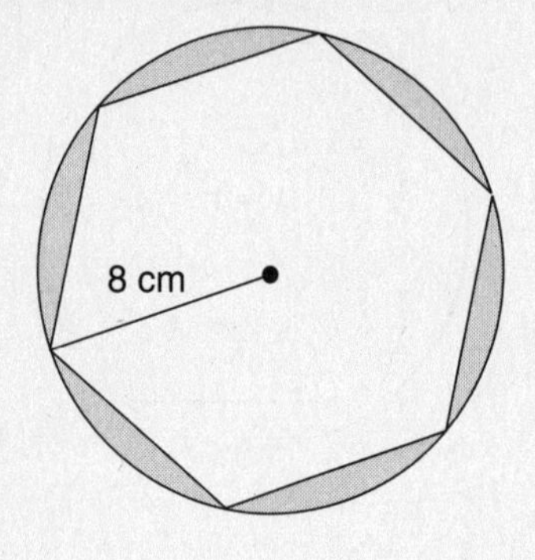

8.

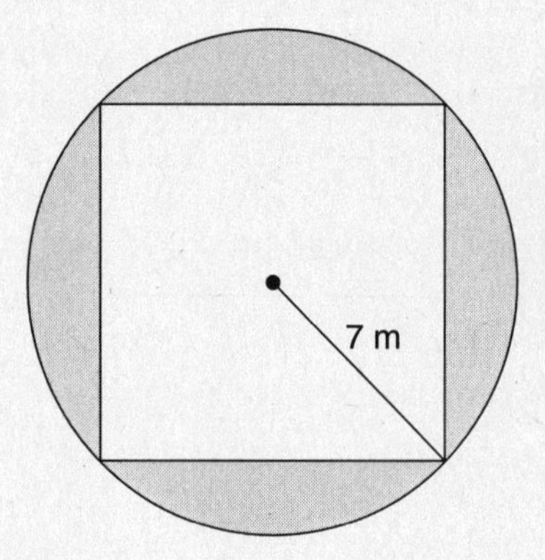

9.

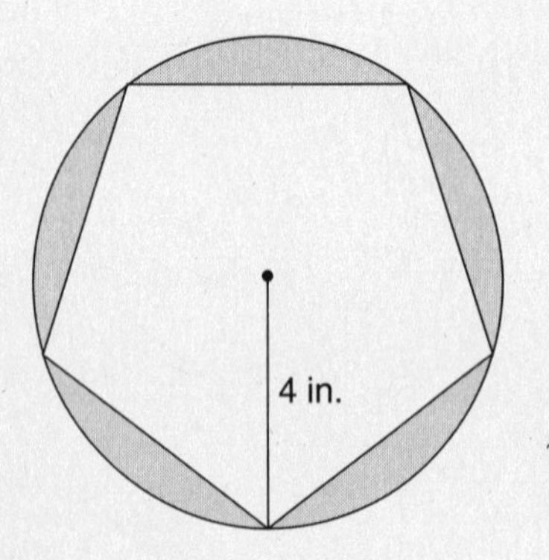

10.

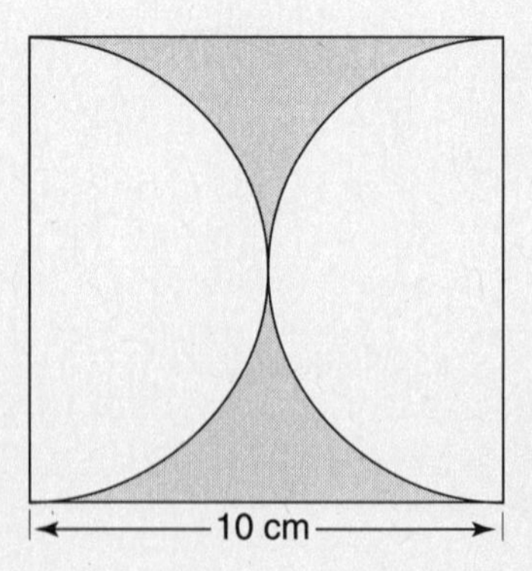

NAME ______________________ DATE ____________

Practice

Integration: Graph Theory Polygons As Networks

Find the degree of each node in the network.

1.

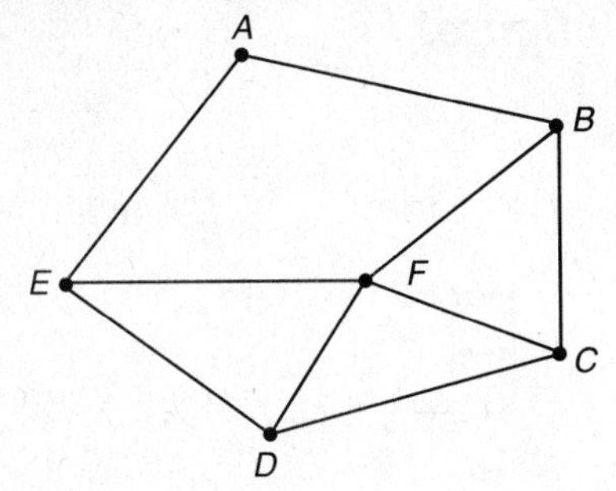

2.

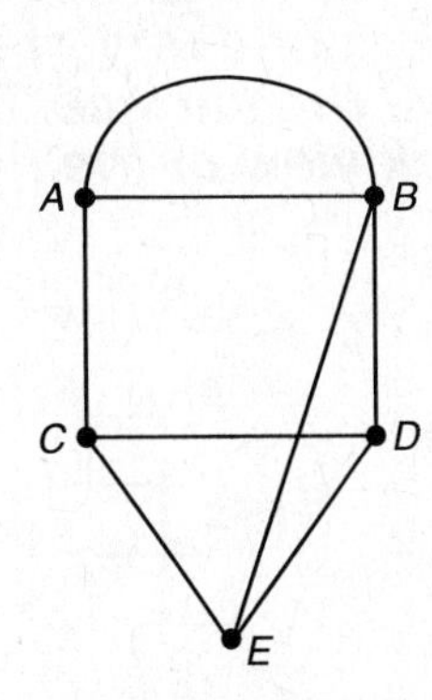

Name the edges that need to be added to make the network complete.

3.

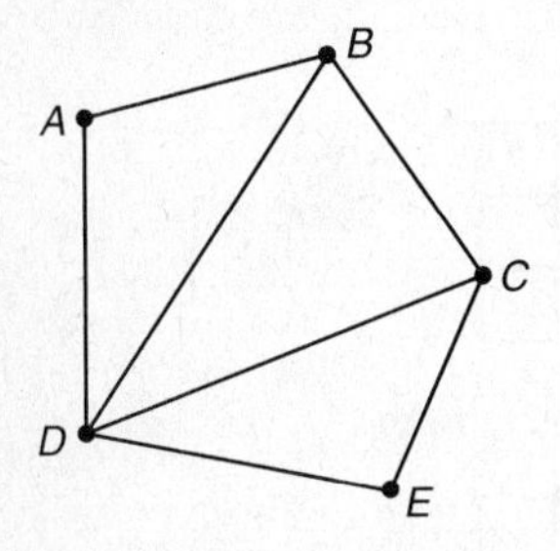

4.

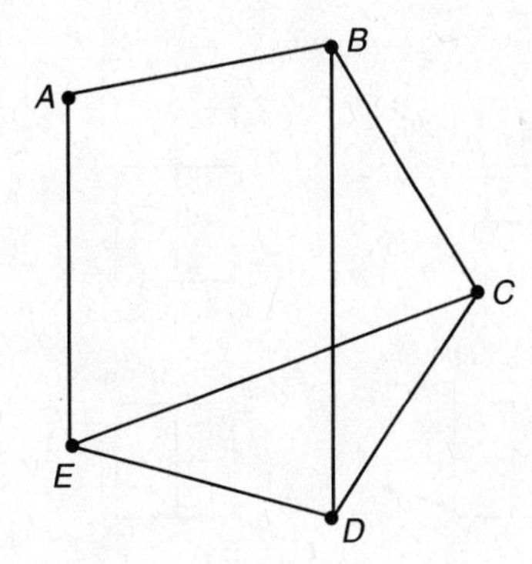

Determine whether each network is traceable. If a network is traceable, use arrows to trace the network.

5.

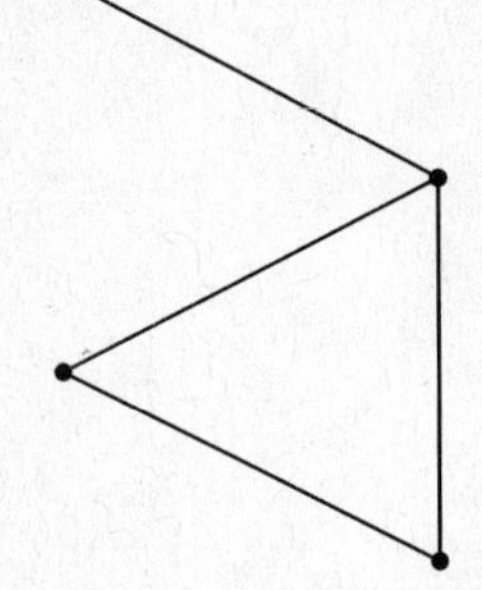

6.

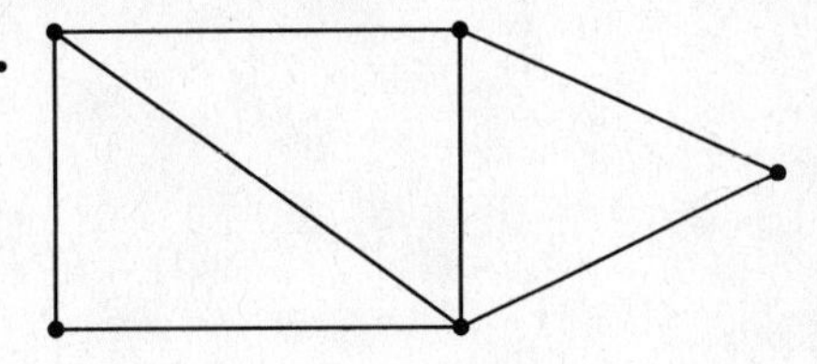

7. Draw two network of your own that are traceable. Then draw two networks that are not traceable.

NAME ________________________ DATE ____________

Practice

Exploring Three-Dimensional Figures

Various views of a solid figure are given below. The edge of one block represents one unit of length. A dark segment indicates a break in the surface. Make a model of each figure. Then draw the back view of the figure.

top view | left view | front view | right view | back view

1.

2.

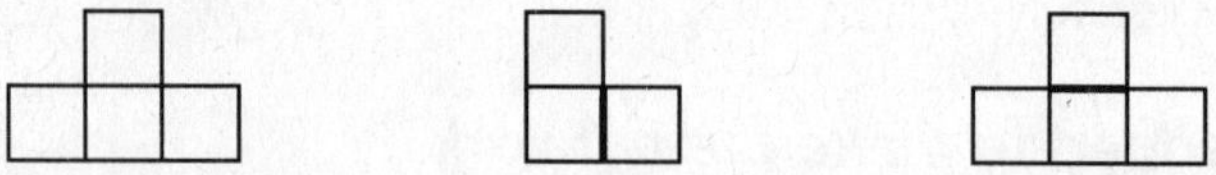

3.

4.

From the views shown in Exercises 1–4, draw a corner view.

5.

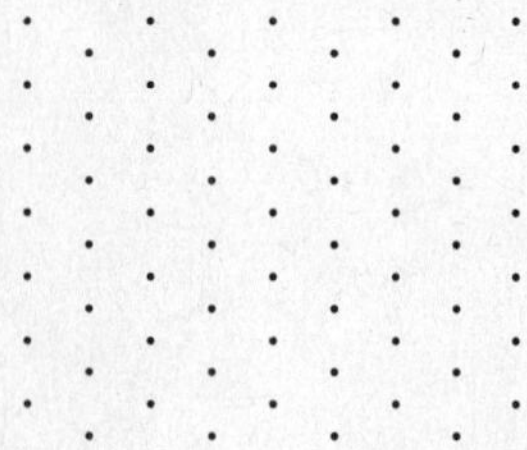

6.

7.

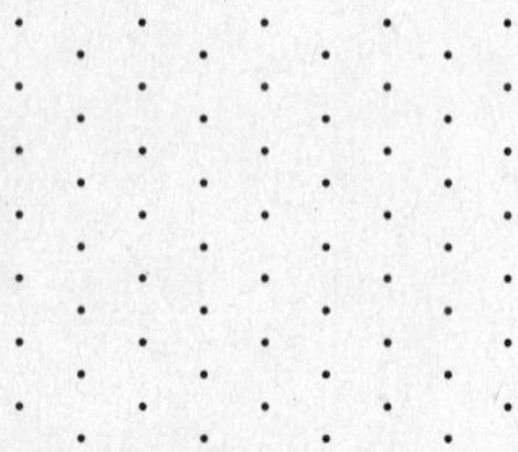

8.

NAME ______________________________ DATE ____________

Practice

Student Edition
Pages 584–589

Nets and Surface Area

Given each polyhedron, label the remaining vertices of its net.

1.

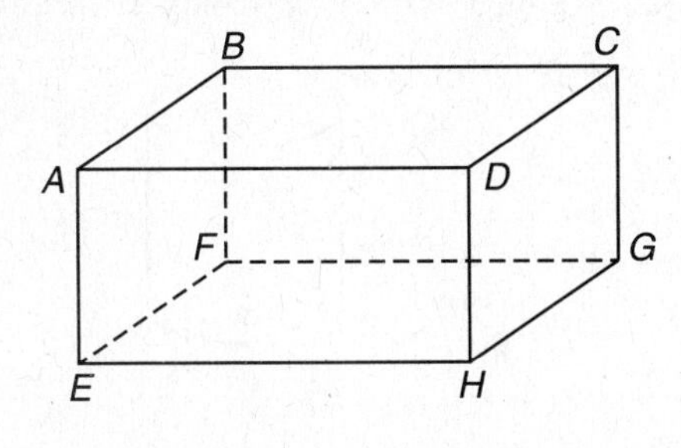

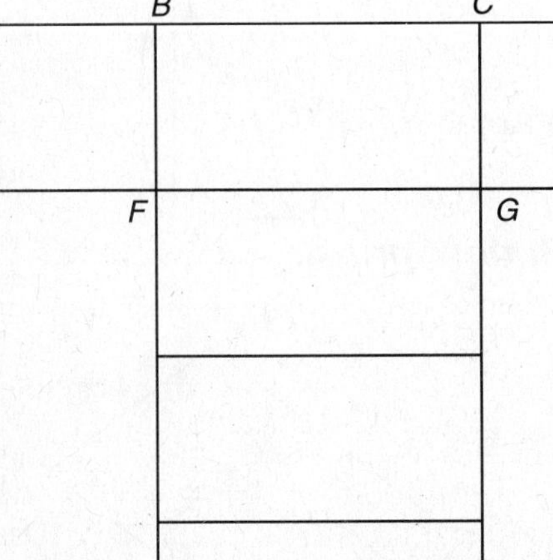

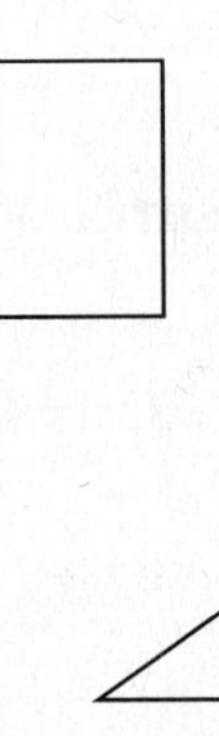

2.

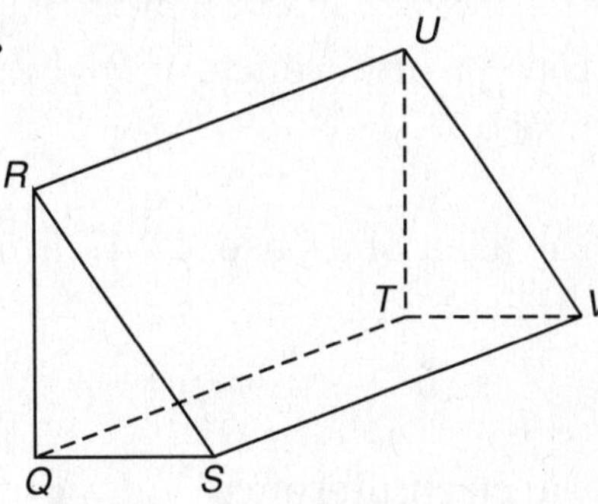

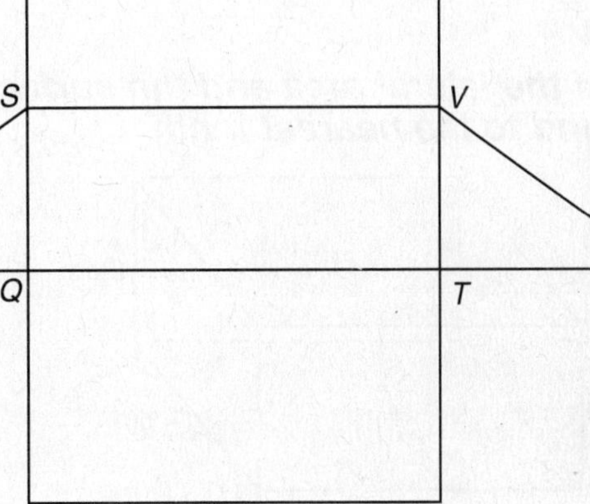

Identify the number and type of polygons that are faces in each polyhedron.

3.

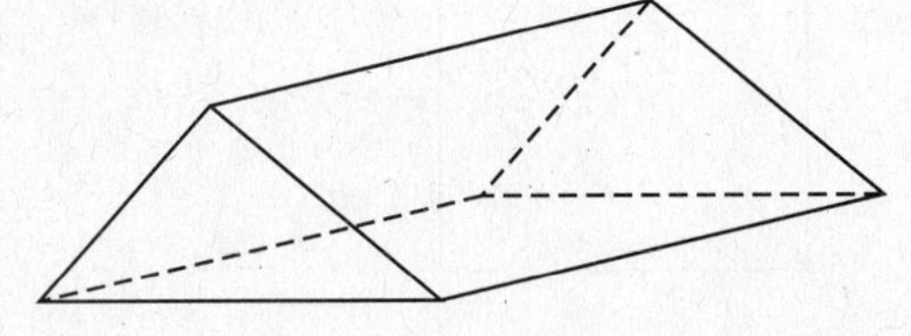

4.

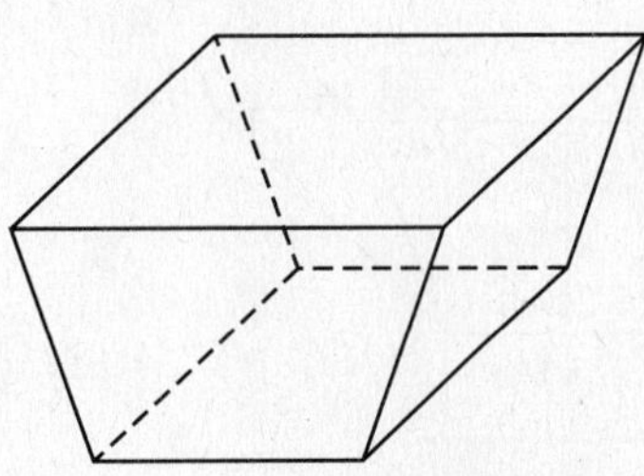

5.

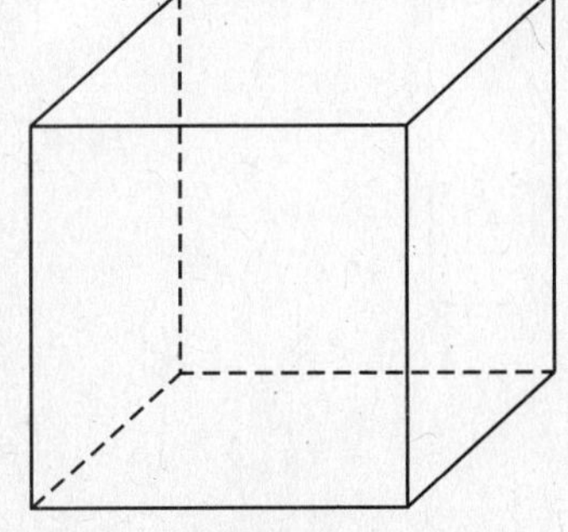

6.

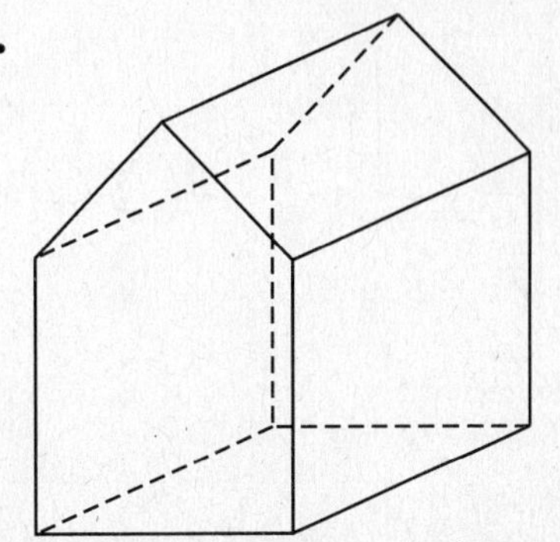

Surface Area of Prisms and Cylinders

Use the right cylinders shown to answer each of the following. Express all answers in terms of π.

1. Find the circumference of the base.

2. Find the lateral area.

3. Find the area of a base.

4. Find the surface area.

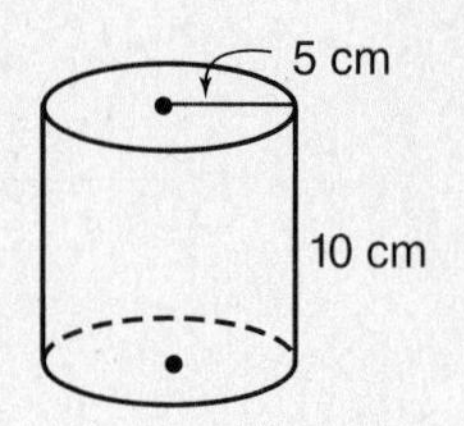

5. Find the circumference of the base.

6. Find the lateral area.

7. Find the area of a base.

8. Find the surface area.

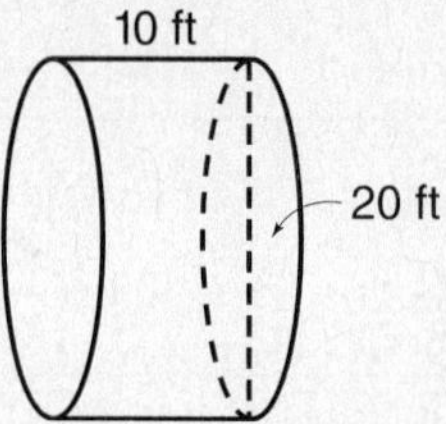

Find the lateral area and the surface area of each right prism. Round to the nearest tenth.

9.

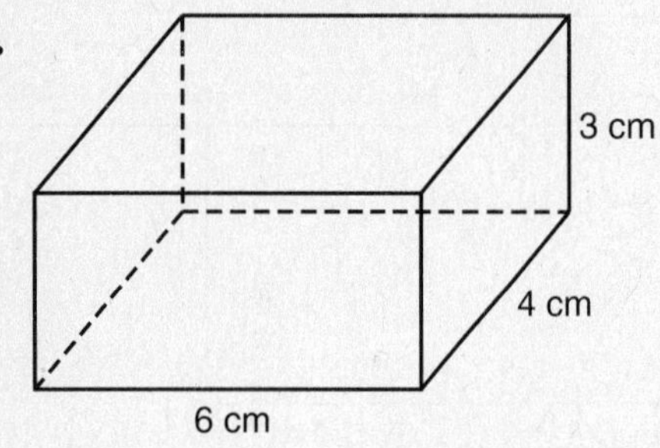

10.

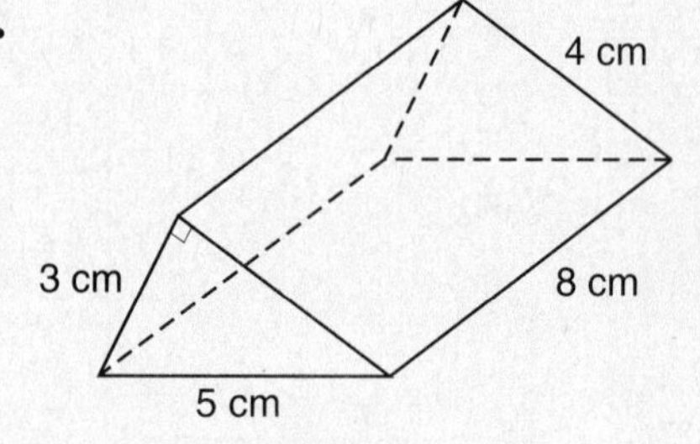

11.

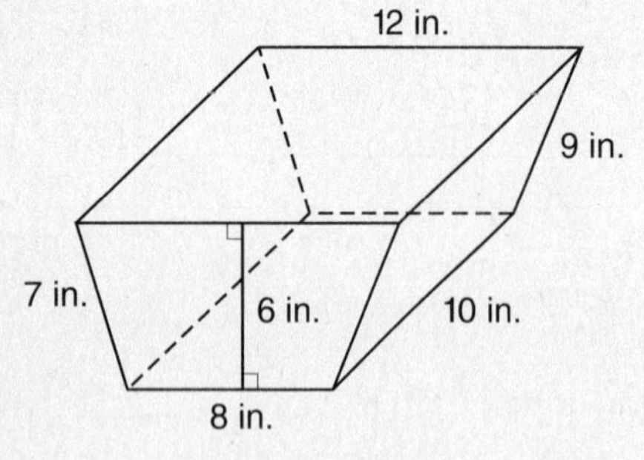

12.

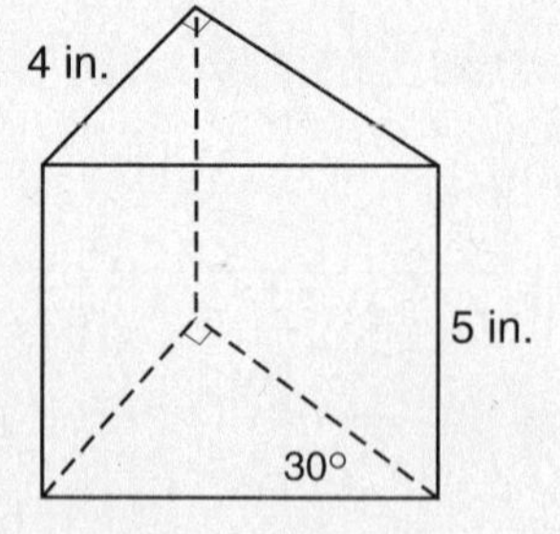

13.

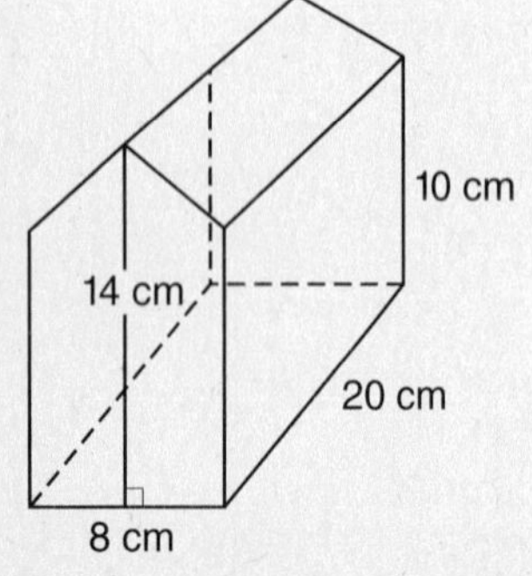

14.

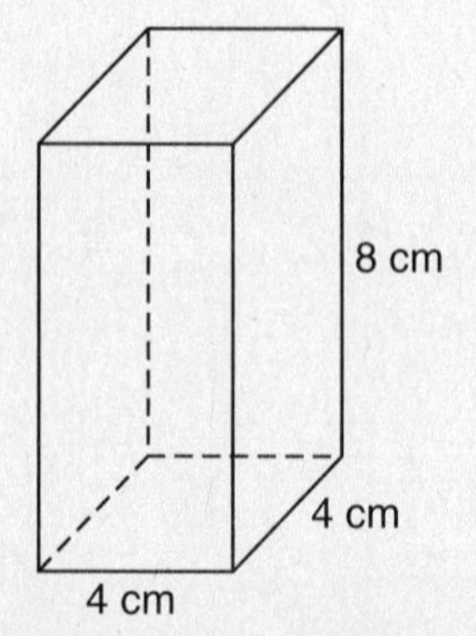

NAME ________________________ DATE ____________

Practice

Surface Area of Pyramids and Cones

Find the lateral area of each regular pyramid or right cone. Round to the nearest tenth.

1.

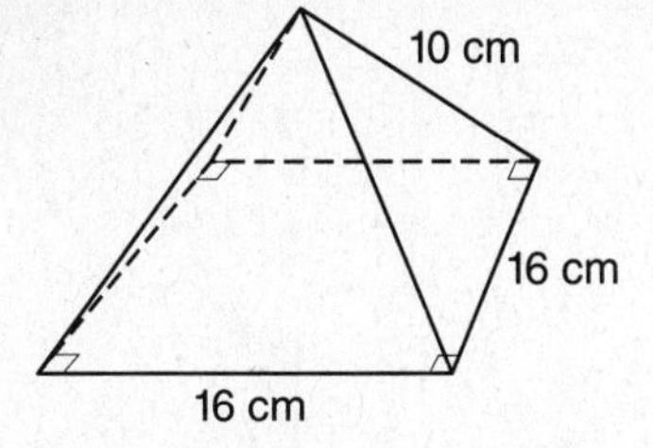

2.

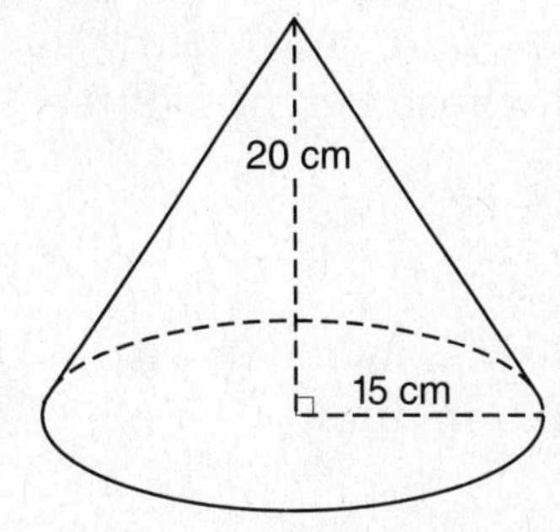

3.

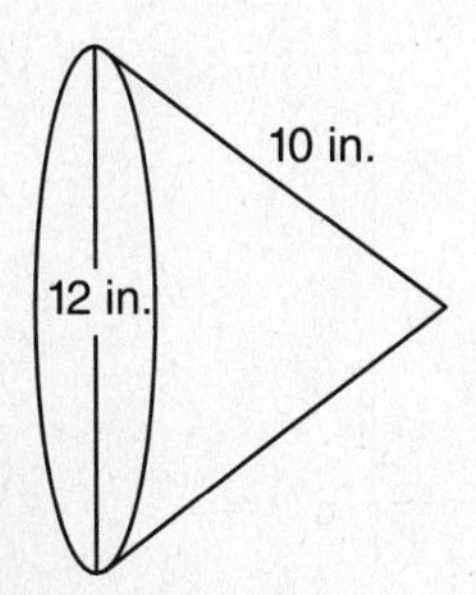

4.

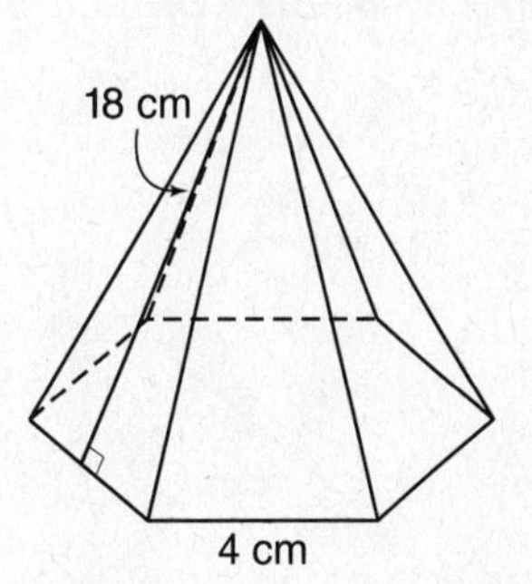

Find the surface area of each solid. Round to the nearest tenth.

5.

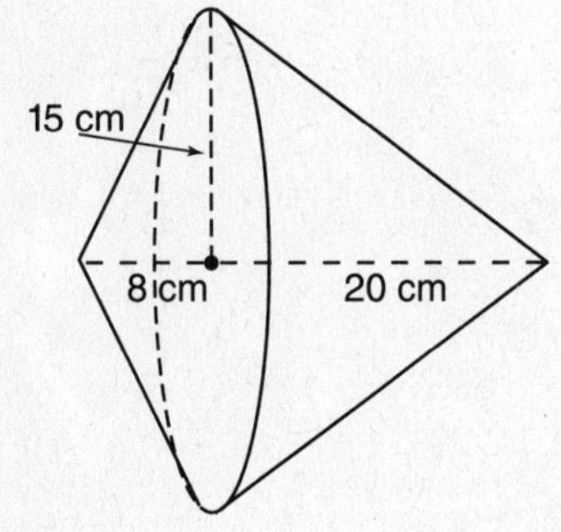

6.

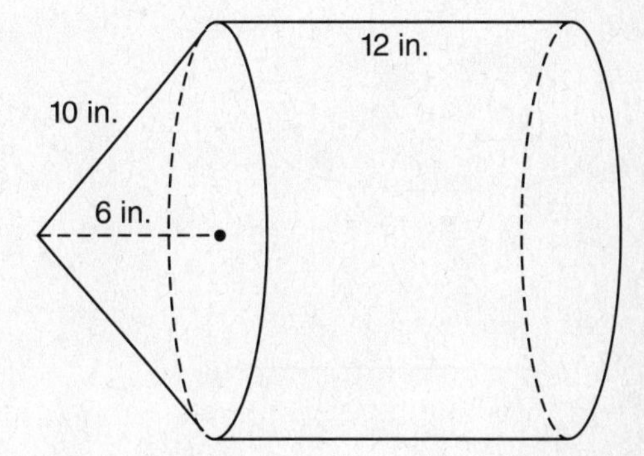

Practice

Volume of Prisms and Cylinders

Find each of the following. Round to the nearest tenth.

1. the volume of a right prism whose square base has sides of 4 feet and whose height is 9 feet

2. the volume of a cylinder with a height of 2 meters and a radius of 0.5 meters

Find the volume of each solid. Round to the nearest tenth.

3.

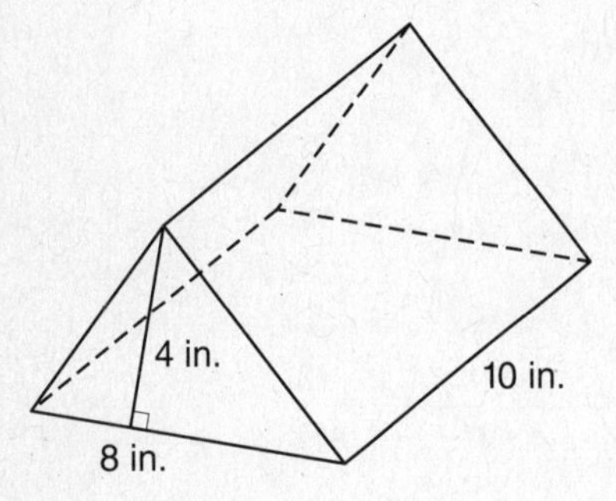

4.

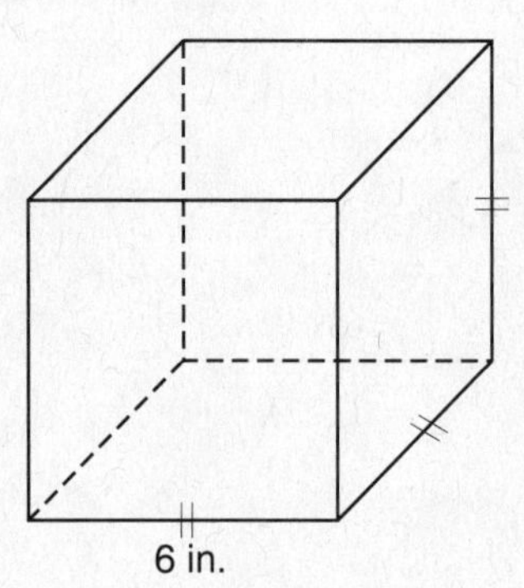

5.

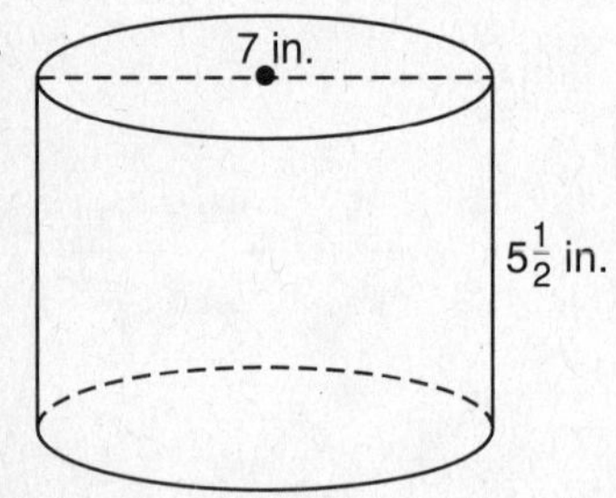

6.

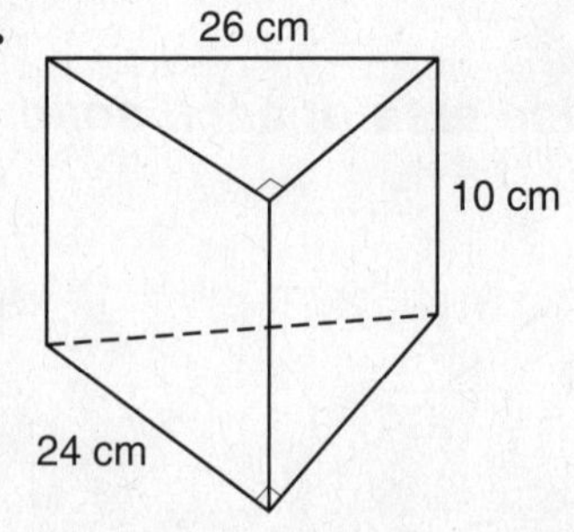

7.

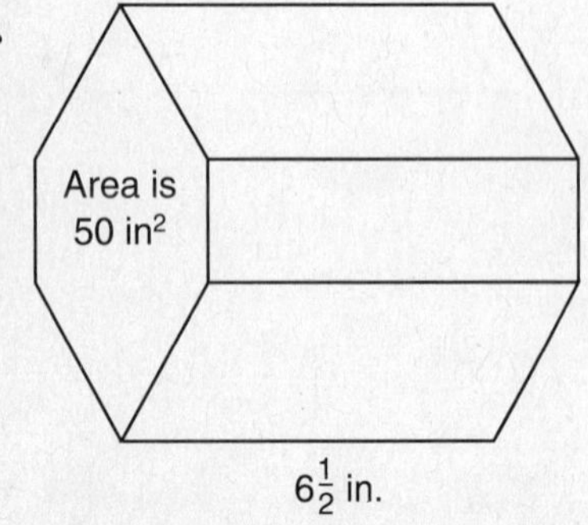

8.

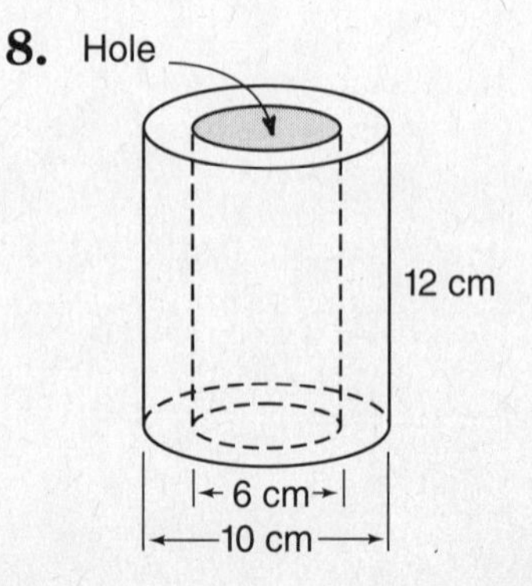

Volume of Pyramids and Cones

Find the volume of each pyramid. Round to the nearest tenth.

1. The base has an area of 84.3 square centimeters, and the height is 16.4 centimeters

2. The base has an area of 17 square feet, and the height is 3 feet.

Find the volume of each cone. Round to the nearest tenth.

3. The base has a radius of 16 centimeters, and the height is 12 centimeters

4. The base has a diameter of 24 meters, and the height is 15.3 meters

Find the volume of each solid. Round to the nearest tenth.

5.

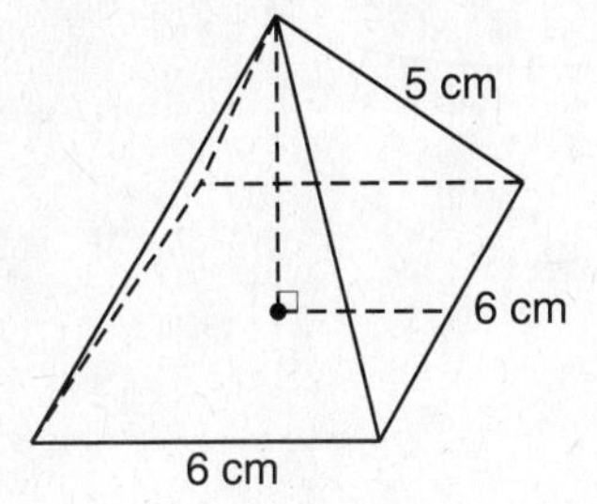

6.

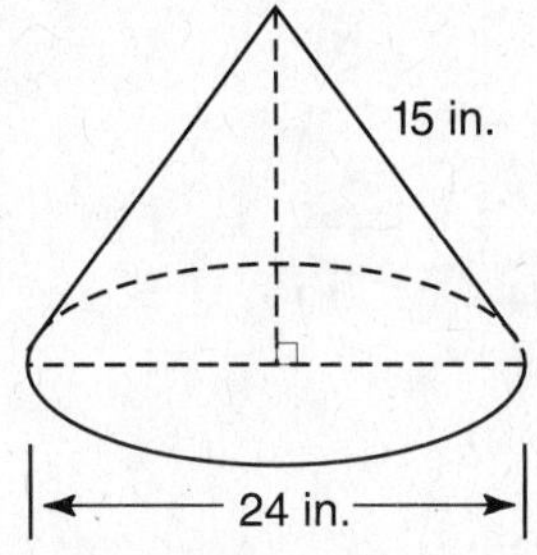

7.

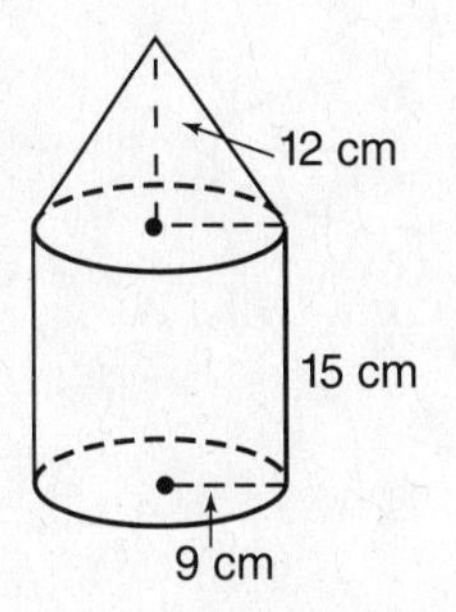

8.

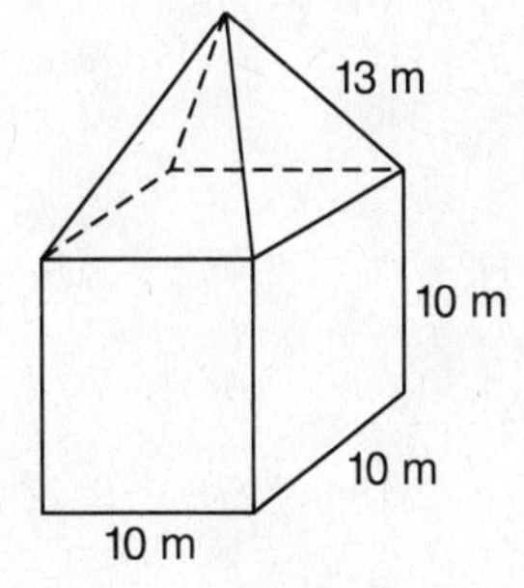

NAME ______________________ DATE __________

Practice

Surface Area and Volume of Spheres

Find the surface area and volume of each sphere described below. Round to the nearest tenth.

1. The diameter is 100 centimeters.

2. A great circle has a circumference 83.92 meters.

3. The radius is 12 inches long.

4. A great circle has an area of 70.58 square feet.

Find the surface area and volume of each solid. Round to the nearest tenth.

5.

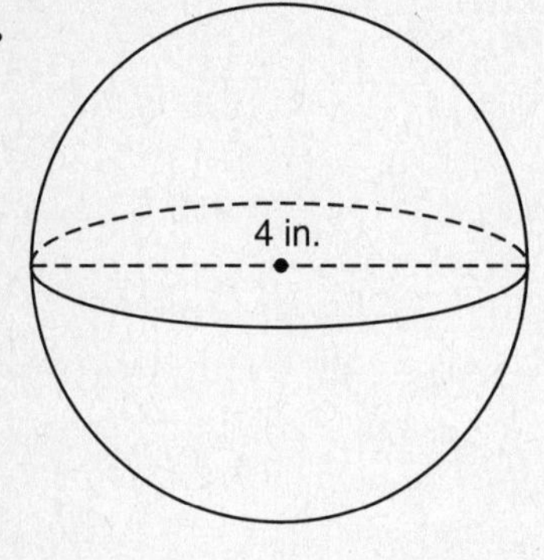

6.

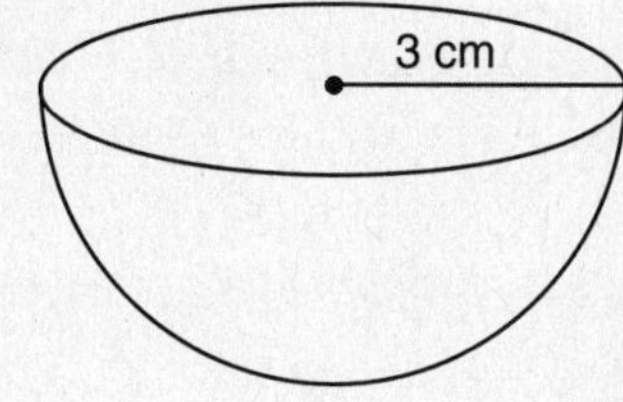

7.

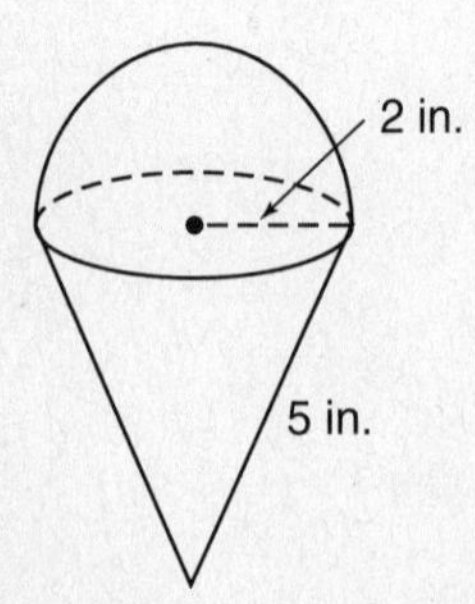

8.

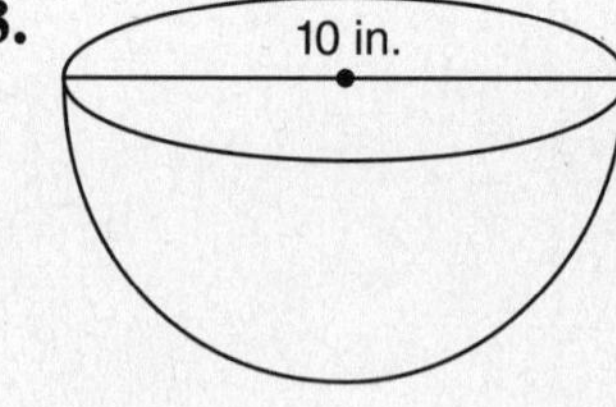

NAME ______________________________ DATE ____________

Practice

Congruent and Similar Solids

Determine if each pair of solids is similar, congruent, or neither.

1.

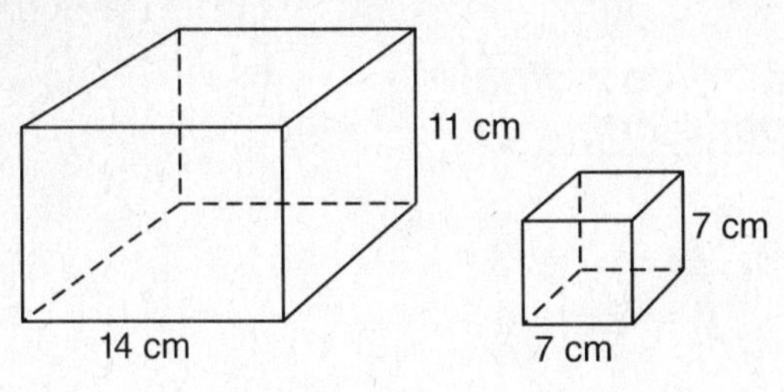

2.

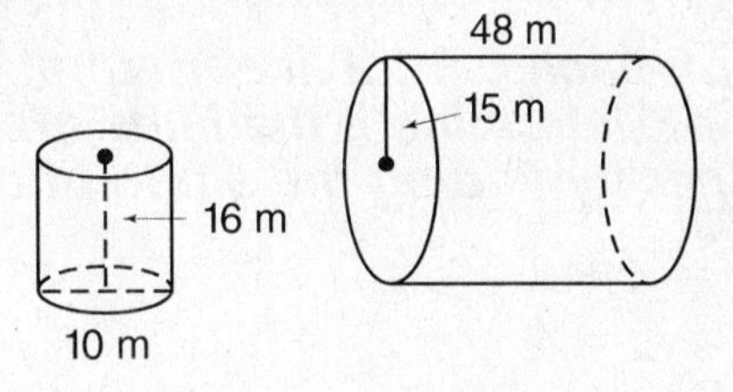

3.

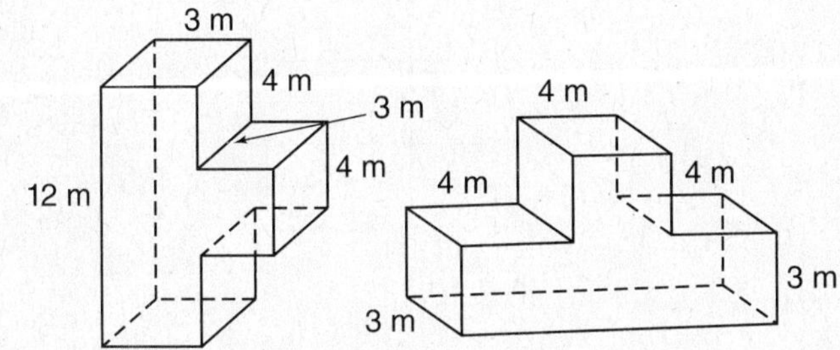

4.

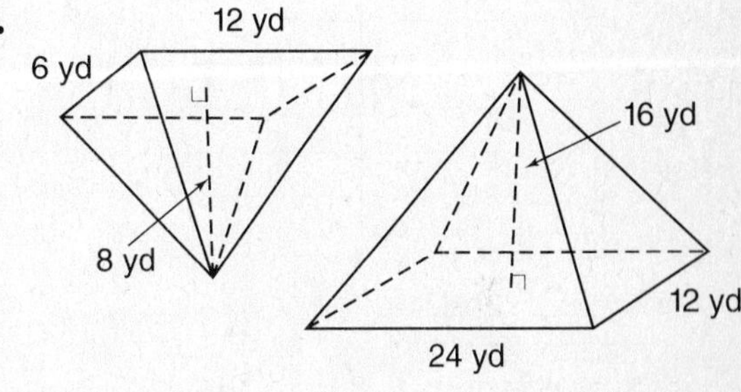

The two right rectangular prisms shown at the right are similar.

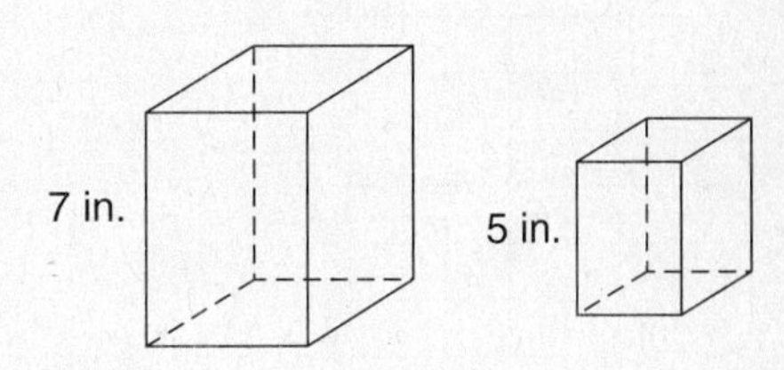

5. Find the ratio of the perimeters of the bases.

6. What is the ratio of the surface areas?

7. Suppose the volume of the smaller prism is 60 in^3.
Find the volume of the larger prism.

Determine if each statement is true or false. If the statement is false, rewrite it so that it is true.

8. If two cylinders are similar, then their volumes are equal.

9. Doubling the height of a cylinder doubles the volume.

10. Two solids are congruent if they have the same shape.

Practice

Integration: Algebra
Graphing Linear Equations

Graph each pair of linear equations on the same coordinate plane. Determine if the lines are parallel, perpendicular, or neither by finding the slope of each line.

1. $y = x + 2$
$y - x = 4$

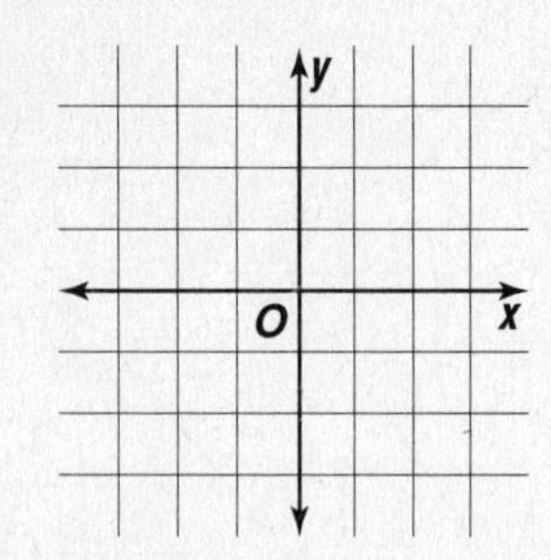

2. $2x - y = 3$
$y = 2x + 4$

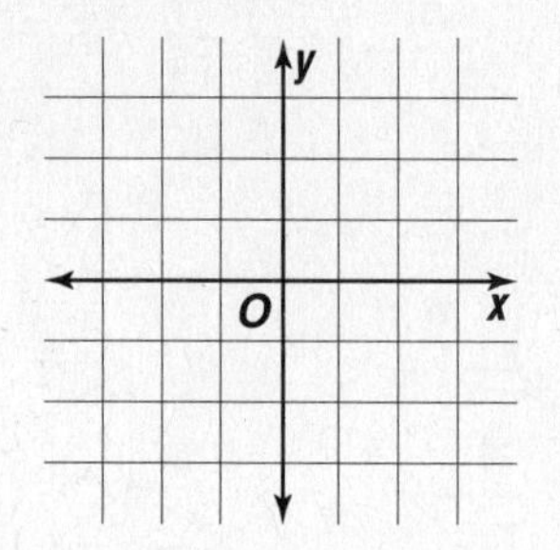

3. $3y = x + 1$
$y = {}^{-}3x + 2$

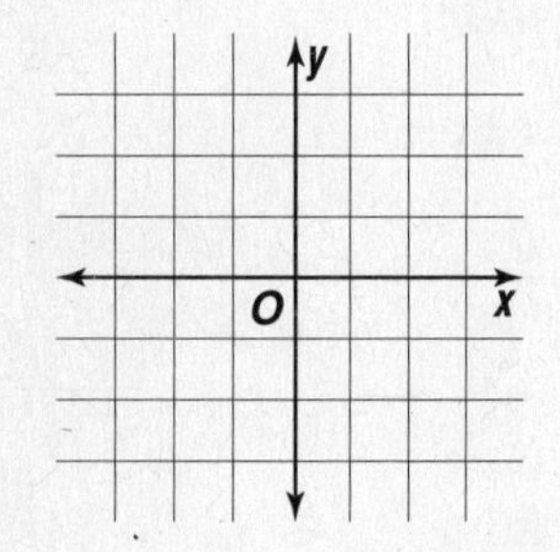

4. $y = 2x + 1$
$y = 3x - 1$

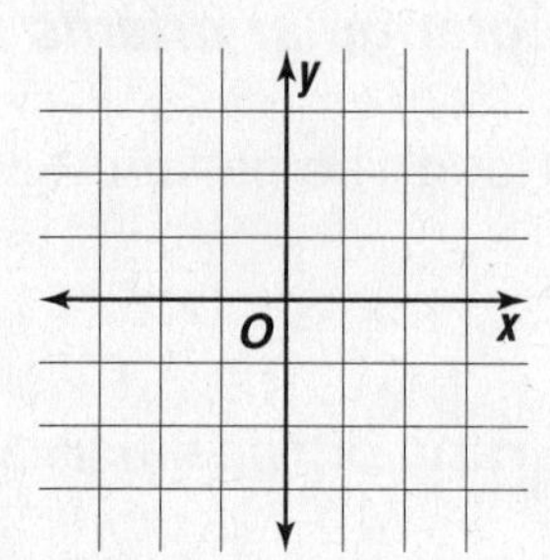

Find the slope and y-intercept of the graph of each equation.

5. $x + y = 8$

6. $2x - y = 4$

7. $2x - 3y = 10$

Determine the x- and y-intercepts of each line. Then graph the equation.

8. $y = 2x - 3$

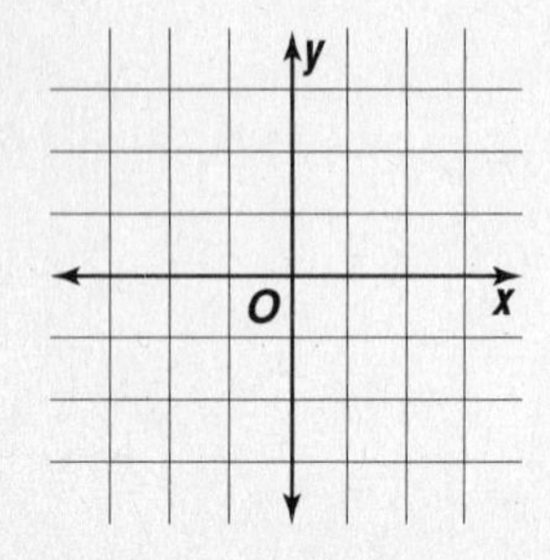

9. $y = 5$

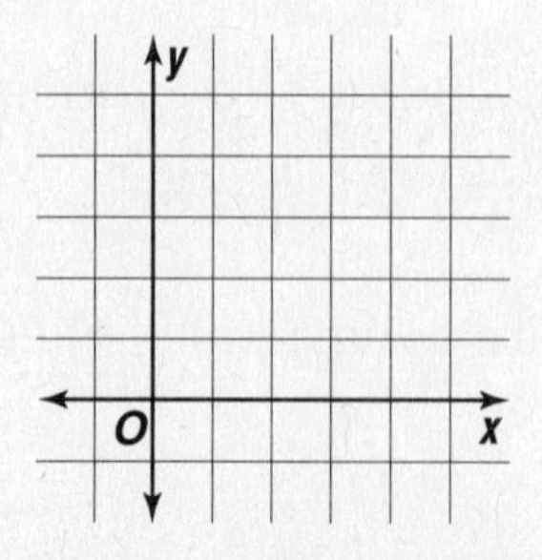

Practice

Integration: Algebra
Writing Equations of Lines

State the slope and y-intercept for each line. Then write the equation of the line in slope-intercept form.

1. $\overleftrightarrow{AB}$

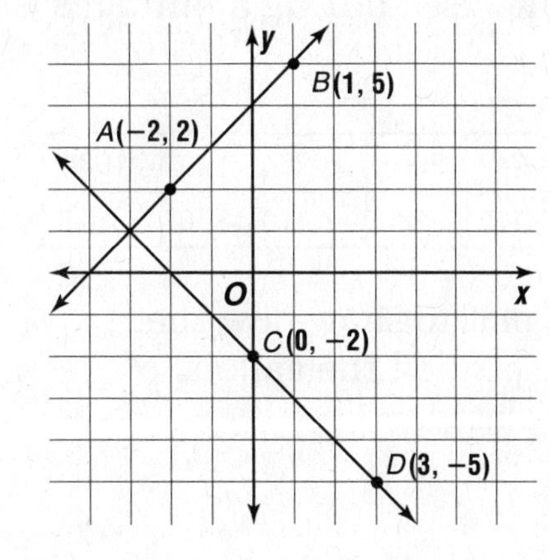

2. $\overleftrightarrow{CD}$

Write the equation in slope-intercept form of the line satisfying the given conditions.

3. parallel to $y = 3x - 2$; y-intercept $= 1$

4. perpendicular to $y = \frac{1}{2}x + 8$; passes through the point at $(0, 3)$

5. perpendicular to the line passing through points at $(-2, 3)$ and $(2, 1)$; y-intercept $= -3$

6. passes through the points at $(1, 5)$ and $(-3, 5)$

7. x-intercept $= 5$; y-intercept $= -2$

8. $m = -\frac{2}{3}$; passes through the point at $(-1, 2)$

9. $m = -4$; x-intercept $= 0$

10. The length of a rectangular garden is 30 feet more than 2 times its width. Its perimeter is 300 feet. Find its length and width.

12–3

NAME ______________________ DATE ______________

Practice

Integration: Algebra and Statistics Scatter Plots and Slope

1. Travel The table below lists the number of gallons of gasoline used during 5 different trips.

Length of trip (in miles)	374	506	2020	144	1034
Gallons of gasoline	17	22	101	6	47

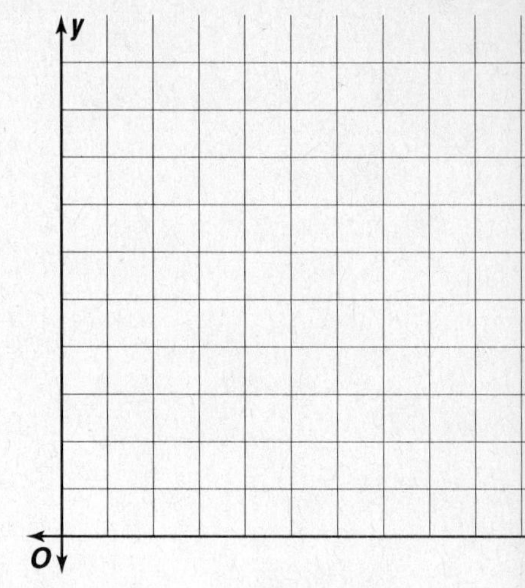

a. Draw a scatter plot to show how the length of the trip x and the gallons of gasoline y are related.

b. Write an equation that relates the trip length to the gallons of gasoline used.

c. About how many gallons of gasoline will be used on an 800-mile trip?

2. Find the equation of the line that is the perpendicular bisector of the segment whose endpoints are $(3, 7)$ and $(^-3, ^-1)$.

3. Find the equations of the lines that contain sides of an isosceles triangle i the vertex of the vertex angle is at th y-intercept of $y = {}^-2x + 3$ and the vertex of a base angle is at $(6, 1)$.

The vertices of △DEF are D(0, 12), E(⁻2, 10), and F(4, 6).

4. Write the equation of the line that contains the altitude to $\overline{EF}$.

5. Write the equation of the line line th contains the perpendicular bisector t $\overline{ED}$.

NAME ____________________ DATE ____________

Practice

Coordinate Proof

Name the missing coordinates in terms of the given variables.

1. $\triangle XYZ$ is isosceles and right.

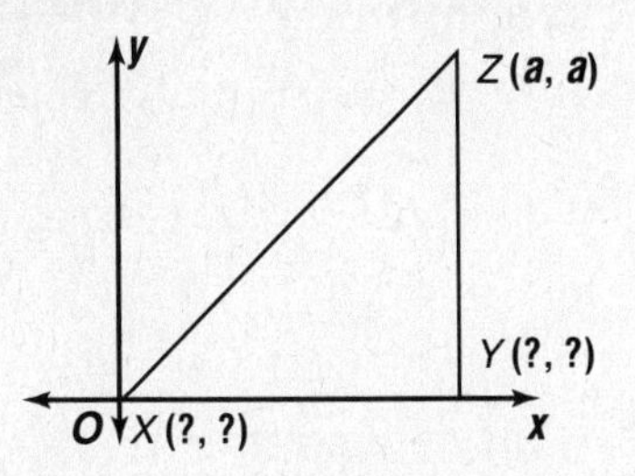

2. *MART* is a rhombus.

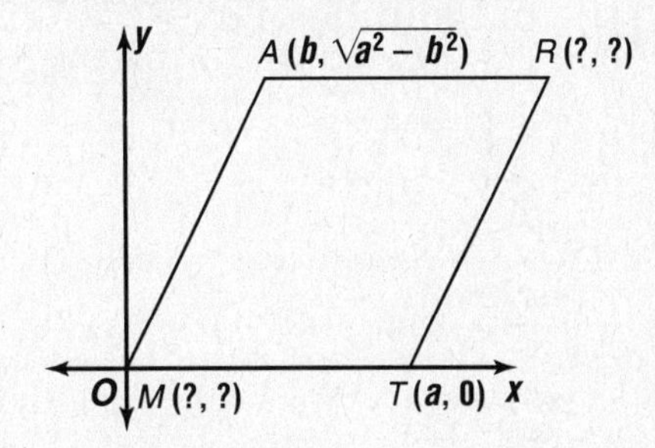

3. *RECT* is a rectangle.

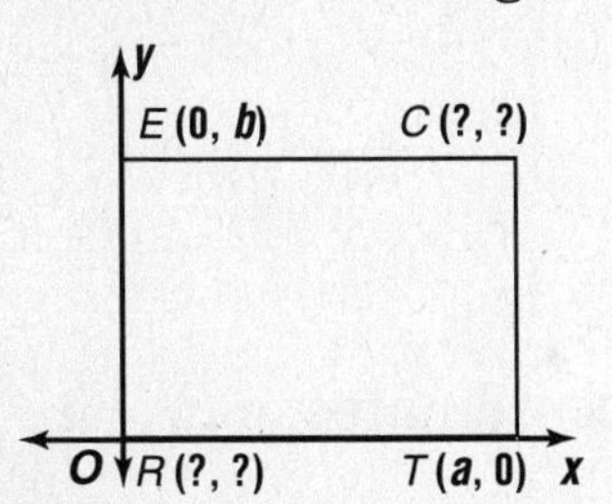

4. *DEFG* is a parallelogram.

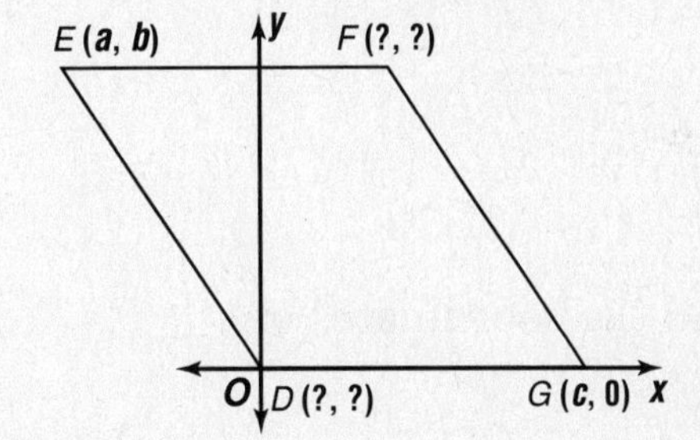

5. Use a coordinate proof to prove that the diagonals of a rhombus are perpendicular. Draw the diagram at the right.

NAME ______________________ DATE ______

Practice

Vectors

Given $\vec{a} = (8, 6)$, $\vec{b} = (4, 3)$, $\vec{c} = (1, 2)$, and $\vec{d} = (-3, -6)$, answer each of the following.

1. Find the magnitude of $\vec{a}$.

2. Find the magnitude $\vec{c}$.

3. Determine if $\vec{b}$ and $\vec{d}$ are equal.

4. Determine if $\vec{c}$ and $\vec{d}$ are equal.

5. Find the coordinates of $\vec{a} + \vec{b}$.

6. Find the coordinates of $(\vec{b} + \vec{c}) + \vec{d}$

7. Given $A(2, 5)$ and $B(7, 10)$, find the magnitude and direction of $\overrightarrow{AB}$.

8. Given $C(0, 1)$ and $D(8, 12)$, find the magnitude and direction of $\overrightarrow{CD}$.

Given path from A south 5 units to B, then east 12 units to C, answer each question.

9. What is the total length of the path?

10. What is the magnitude of $\overrightarrow{AC}$?

NAME ______________________ DATE ____________

Practice

Coordinates in Space

Determine the distance between each pair of points.

1. $A(0, 0, 0)$ and $B(1, 2, 3)$

2. $C(4, -2, 3)$ and $D(0, 2, 0)$

3. $E(1, -2, 5)$ and $F(1, 2, 5)$

4. $P(0, 1, 0)$ and $Q(-1, 0, 1)$

Determine the coordinates of the midpoint of each line segment whose endpoints are given.

5. $A(0, 0, 4)$, $B(4, -6, 6)$

6. $C(-1, 2, 4)$, $D(3, -6, 8)$

7. $E(-2, -3, 6)$, $F(0, -6, 8)$

8. $G\left(-1, 5, \frac{3}{2}\right)$, $H\left(1, -5, \frac{1}{2}\right)$

Write an equation of the sphere given the coordinates of the center and the measure of the radius.

9. $C(0, -3, 1)$, $r = 3$

10. $C(-2, 1, 3)$, $r = 1\frac{1}{2}$

11. $C(3, 0, -1)$, $r = \frac{\sqrt{2}}{2}$

12. $C(5, 5, 5)$, $r = 7$

13. Find the perimeter of a triangle whose vertices are $A(0, 2, 1)$, $B(-2, 2, 6)$, and $C(4, 2, -2)$.

The diameter of a sphere has endpoints $A(-3, 2, 4)$ and $B(1, -6, 5)$.

14. Determine the center of the sphere.

15. Determine the radius of the sphere.

16. Write the equation of the sphere.

17. Find the surface area of the sphere.

Practice

What Is Locus?

Draw a figure and describe the locus of points that satisfy each set of conditions.

1. all points in a plane that are midpoints of the radii of a given circle

2. all points in a plane that are equidistant from the endpoints of a given segment

3. all points in a plane that are equidistant from two parallel lines 12 centimeters apart

4. all points in a plane that are 4 centimeters away from $\overleftrightarrow{AB}$

5. all points in a plane that are equidistant from two concentric circles whose radii are 8 inches and 12 inches

6. all points in a plane that are centers of circles having a given line segment as a chord

7. all points in a plane that belong to a given angle or its interior and are equidistant from the sides of the given angle

NAME ____________________ DATE __________

Practice

Locus and Systems of Equations

Graph each pair of equations to find the locus of points that satisfy both equations.

1. $x + y = 3$
$x - y = 1$

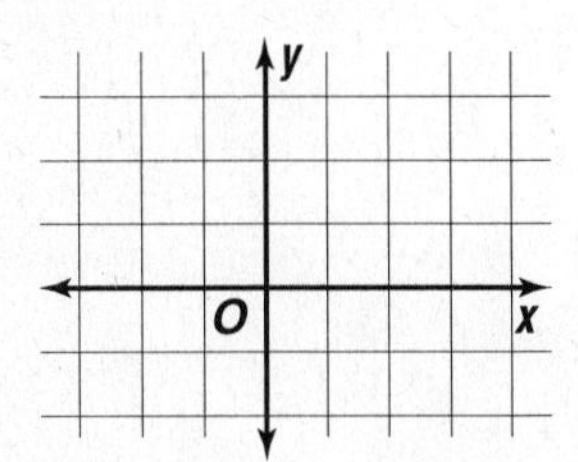

2. $y = 3 + x$
$x + y = 5$

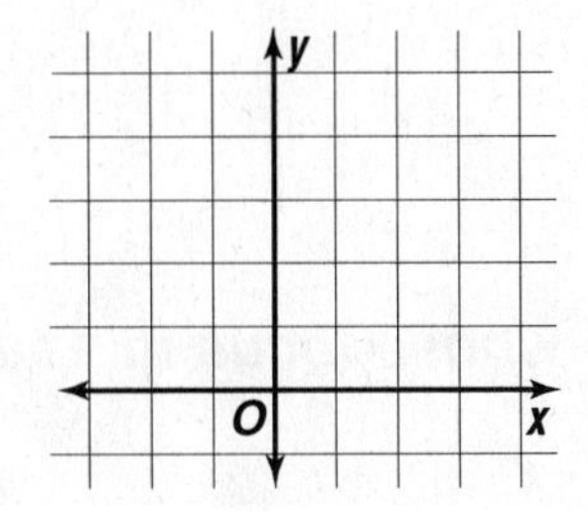

3. $y = 2x$
$x + y = 3$

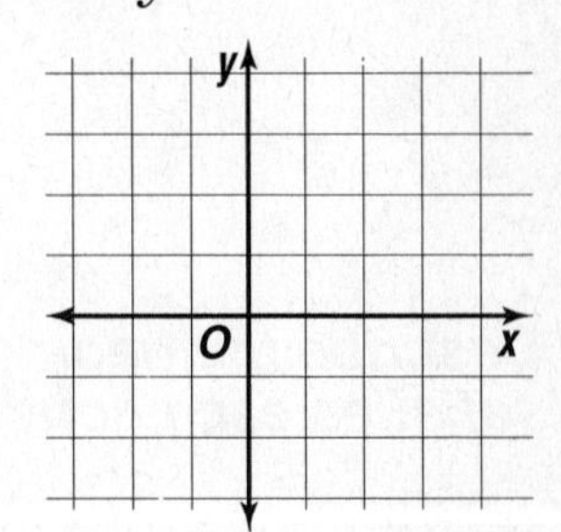

Use either substitution or elimination to find the locus of points that satisfy both equations.

4. $x + y = 7$
$x - y = 9$

5. $x + y = 3$
$3x - 5y = 17$

6. $y = 2x$
$3x + y = 5$

7. $4x - 3y = -1$
$x + 1 = y$

8. $2x + 3y = -1$
$3x + 5y = -2$

9. $3y = 2 - x$
$2x = 7 - 3y$

10. $y = 2x + 1$
$y = 4x + 7$

11. $x = 4$
$y = 3x - 5$

12. $3x + 2y = 10$
$6x - 3y = 6$

Practice

Intersection of Loci

Describe the locus of points in a plane that satisfy each condition.

1. $x + y = 4$

2. $x = y$

3. $y = 5$

4. $x^2 + (y - 2)^2 = 25$

Describe the geometric figure whose locus in space satisfies each condition.

5. $(x + 2)^2 + (y - 3)^2 + z^2 = 64$

6. $(x - 2)^2 (y + 3)^2 + (z - 1)^2 = 81$

Draw a diagram and describe the locus of points.

7. all points in a plane that are 3 inches from a given segment and equidistant from the two endpoints

8. all points in a plane that are 3 inches from a given line and 3 inches from a given point on the line

9. all points in a plane that are equidistant from the vertices of a given square

10. all points in a plane that are 5 centimeters from a given point A and equidistant from points A and B that are 8 centimeters apart

NAME________________________ DATE ____________

Practice

Student Edition
Pages 715–721

Mappings

Each figure below has a preimage or is the image of isometry. Write the image of each given preimage listed in Exercise 1–6.

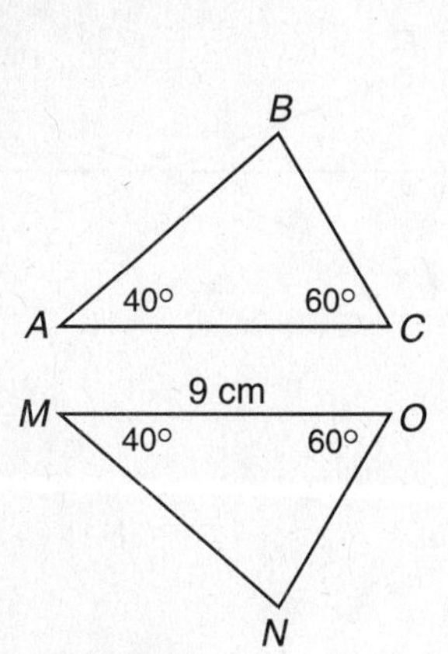

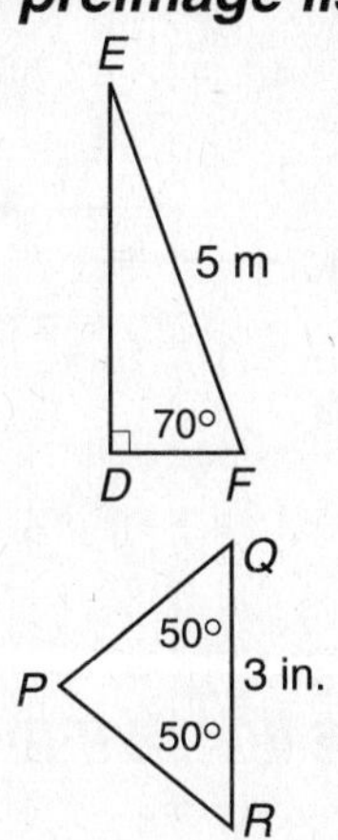

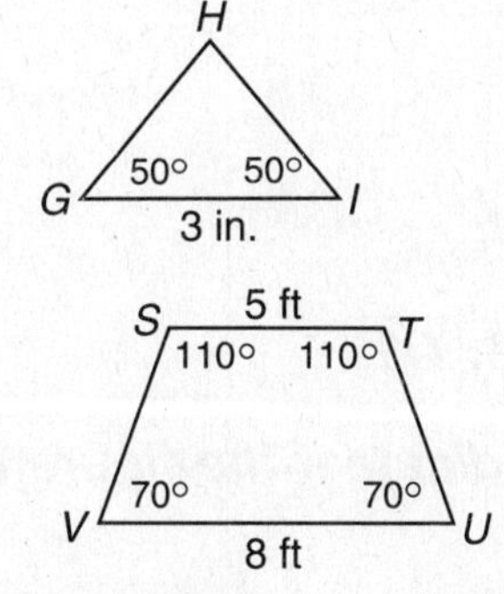

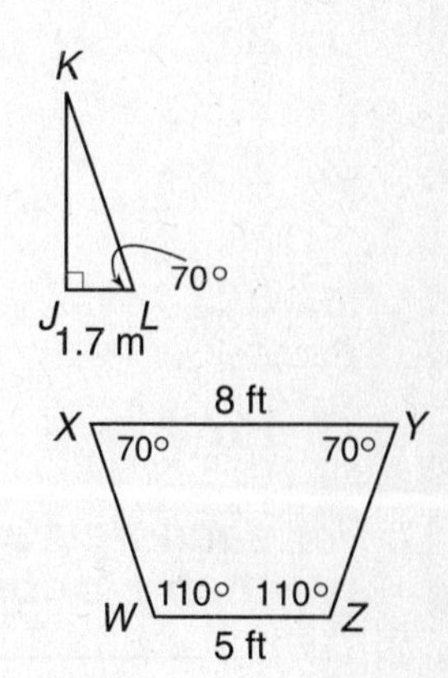

1. $\triangle ABC$

2. $\triangle DEF$

3. $\triangle GHI$

4. $\triangle ONM$

5. Quadrilateral *STUV*

6. Quadrilateral *XYZW*

7. Make a Table Benjamin has seven employees who work in his store: Janell, Pedro, David, Elyse, Faye, Gordon, and Irene. Janell and Gordon work full time, 5 days in a row per week. Pedro and Faye each work 4 days per week. Pedro cannot work on Thursdays. David and Elyse also work 4 days per week each. David cannot work on weekends, and Elyse cannot work on Wednesdays. Irene works any three days assigned each week. Benjamin needs three people to work on Monday through Thursday, five to work on Friday, and six to work on weekends. Make a possible schedule for Benjamin to use.

13-5

NAME ______________________ DATE ______

Practice

Student Editi
Pages 722–7

Reflections

For the figure at the right, name the reflection image of each of the following over line ℓ.

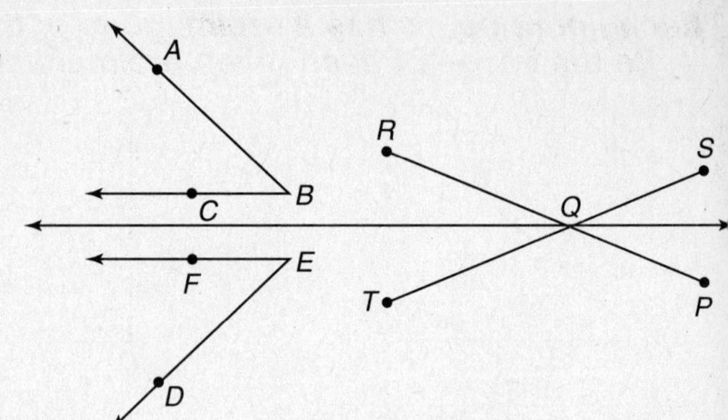

1. A **2.** C

3. T **4.** Q

4. $\overline{BC}$ **6.** $\angle CBA$

5. $\overline{PR}$ **8.** $\overline{QT}$

For each figure, indicate if the figure has line symmetry, point symmetry, or both.

9.

10.

11.

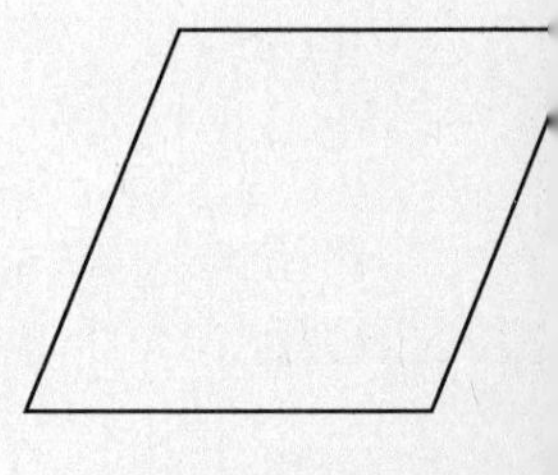

Use a straightedge to draw the reflection image of each figure over line m.

12.

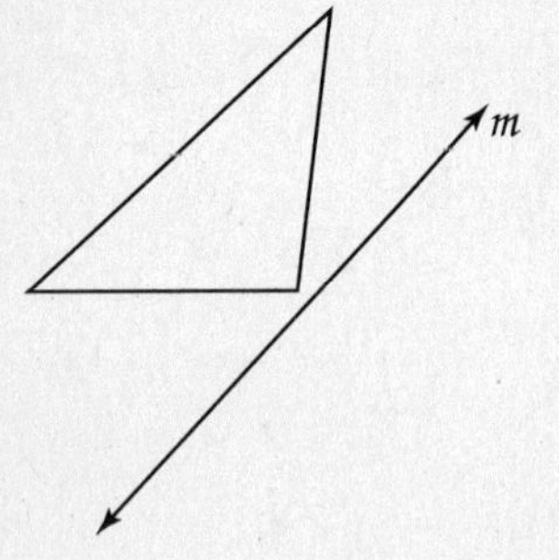

13.

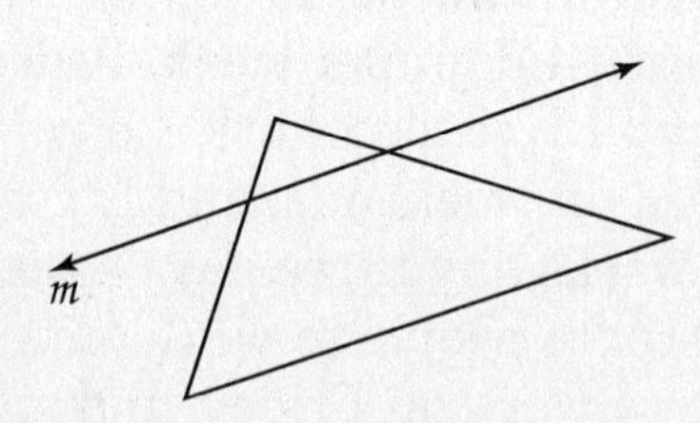

14.

Draw all possible lines of symmetry. If none exist, write none.

15.

16.

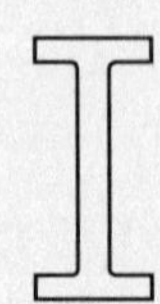

17.

Practice

Translations

For each of the following, lines ℓ and m are parallel. Determine whether Figure 3 is a translation image of Figure 1. Write yes or no. Explain your answer.

1.

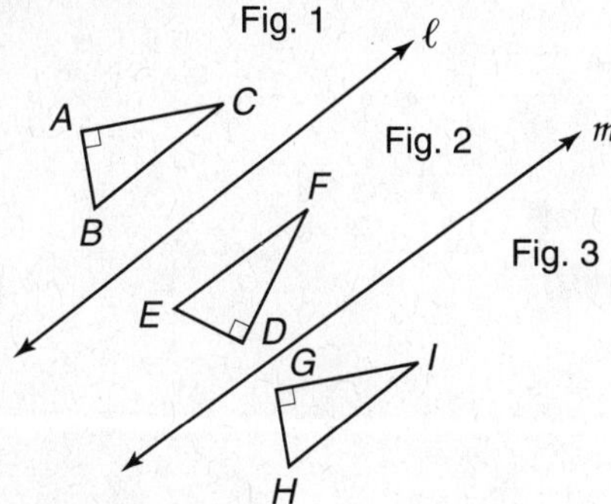

2.

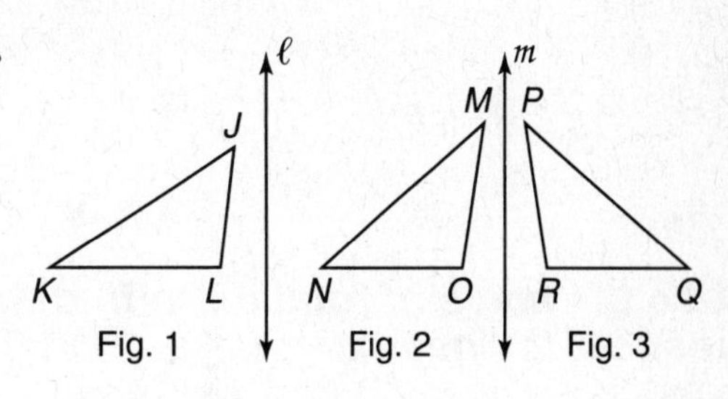

3.

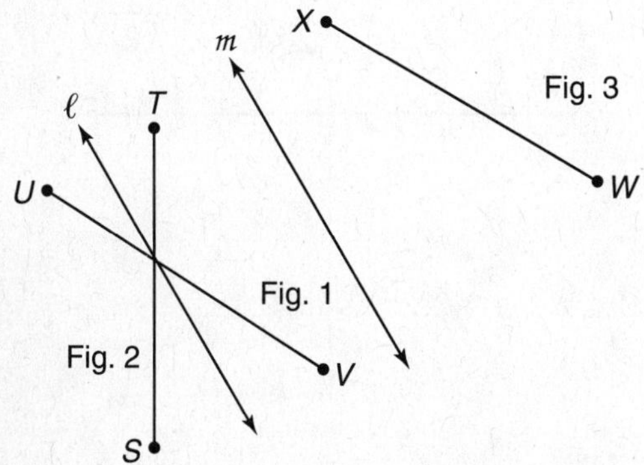

4.

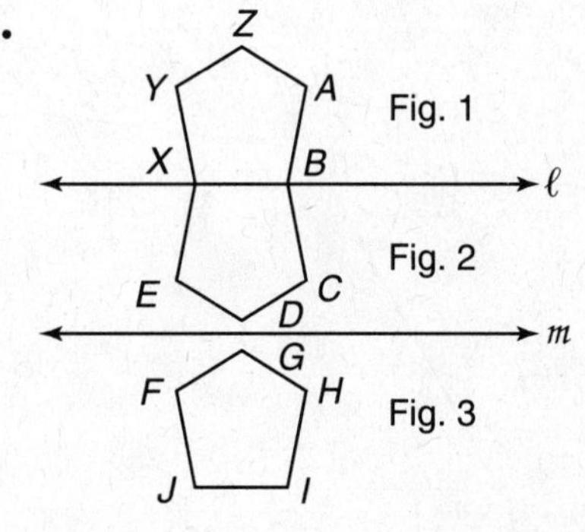

For Exercises 5–8, refer to the figures above.

5. Name the reflection of $\overline{ST}$ with respect to line ℓ. If none is drawn, write *none*.

6. Name the reflection of $\triangle JKL$ with respect to line ℓ. If none is drawn, write *none*.

7. Name the reflection of pentagon $XYZAB$ with respect to line ℓ. If none is drawn, write *none*.

8. Name the reflection of $\overline{UV}$ with respect to line m. If none is drawn, write *none*.

Find the translation image of each geometric figure with respect to the parallel lines ℓ and m.

9.

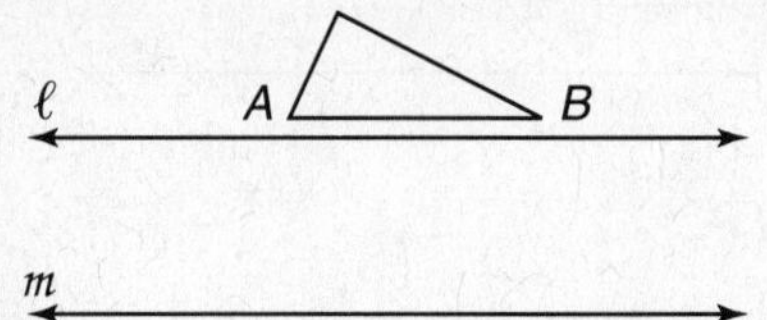

10.

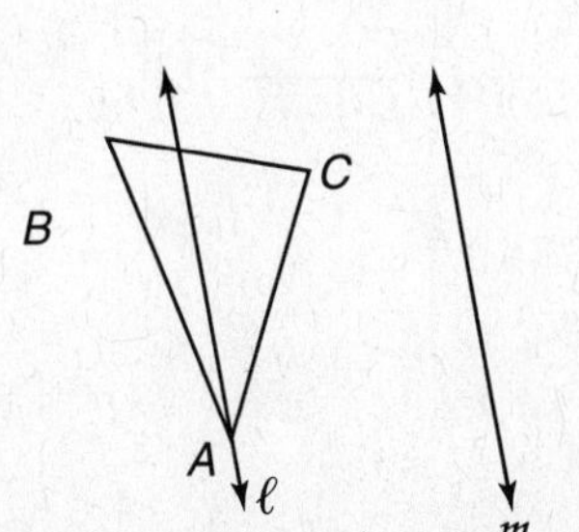

Practice

Rotations

Two lines intersect to form an angle with the following measure. Find the angle of rotation for each.

1. 40° **2.** 72° **3.** 23°

4. 35° **5.** 55° **6.** 63°

Use the angle of rotation to find the rotation image with respect to lines s and t.

7.

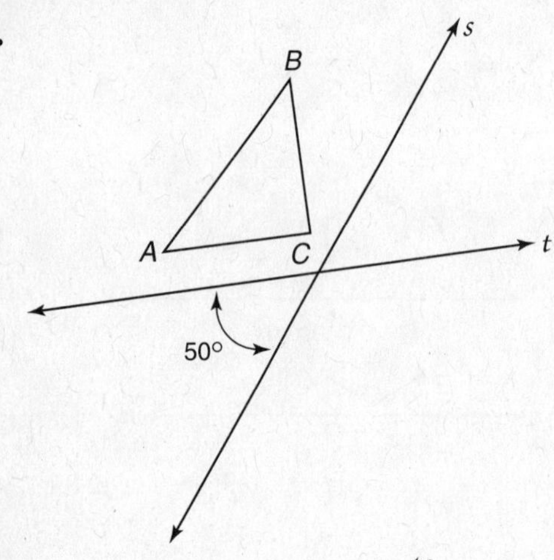

8.

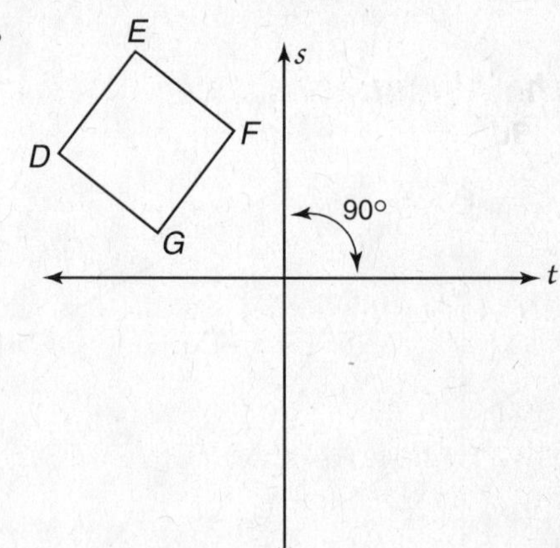

9.

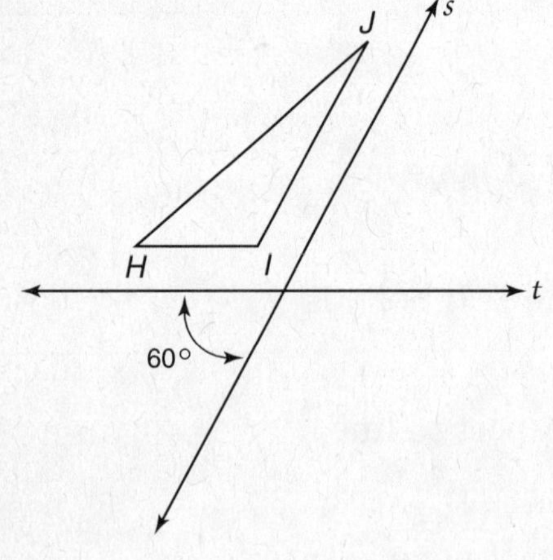

10.

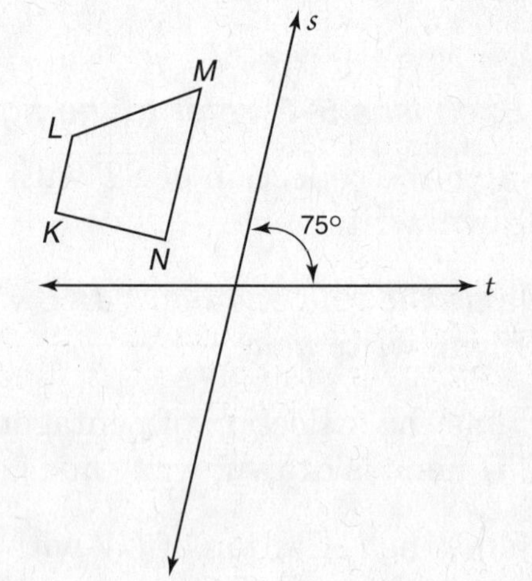

11.

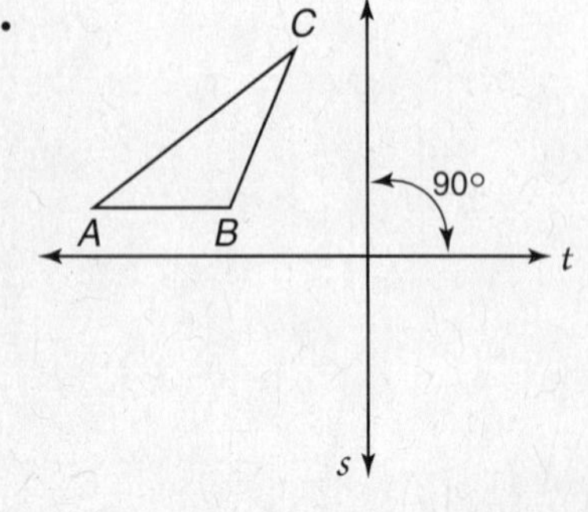

12.

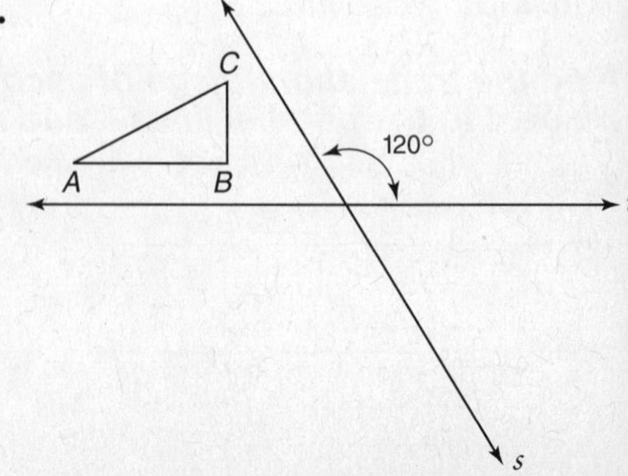

NAME______________________________ DATE ____________

Practice

Dilations

A dilation with center C and a scale factor k maps X onto Y. Find $|k|$ for each dilation. Then determine whether each dilation is an enlargement, a reduction, or a congruence transformation.

1. $CY = 15, CX = 10$

2. $CY = 2, CX = 2$

3. $CX = 5, CY = 2$

4. $CY = 20, CX = \frac{1}{2}$

Find the measure of the dilation image of $\overline{AB}$ with the given scale factor.

5. $AB = 6$ in., $k = -\frac{2}{3}$

6. $AB = 4$ in., $k = 1$

7. $AB = 1\frac{1}{2}$ in., $k = \frac{1}{2}$

8. $AB = 20$ in., $k = -2\frac{1}{2}$

Find each scale factor, find the image of A with respect to a dilation with center C.

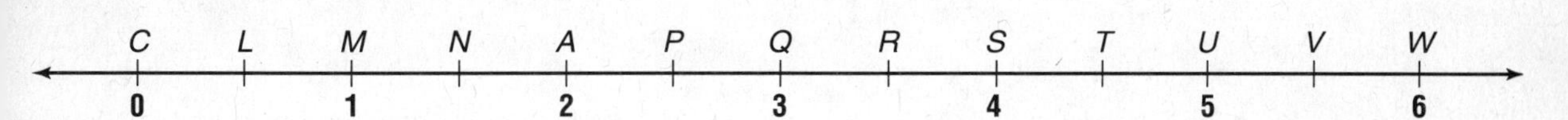

9. 3

10. $\frac{1}{4}$

11. $2\frac{1}{4}$

12. $\frac{3}{4}$

Graph each set of ordered pairs. Then connect the points in order. Using (0, 0) as the center of dilation and a scale factor of 2, draw the dilation image. Repeat this using a scale factor of $\frac{1}{2}$.

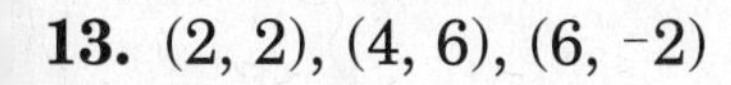

13. (2, 2), (4, 6), (6, −2)

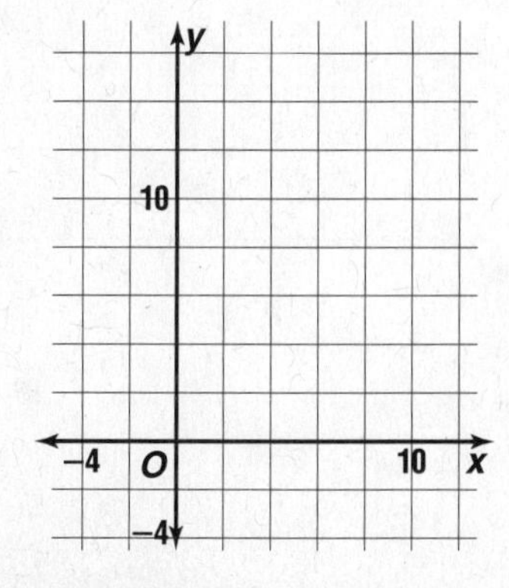

14. (0, 2), (−4, 2), (−4, −2)

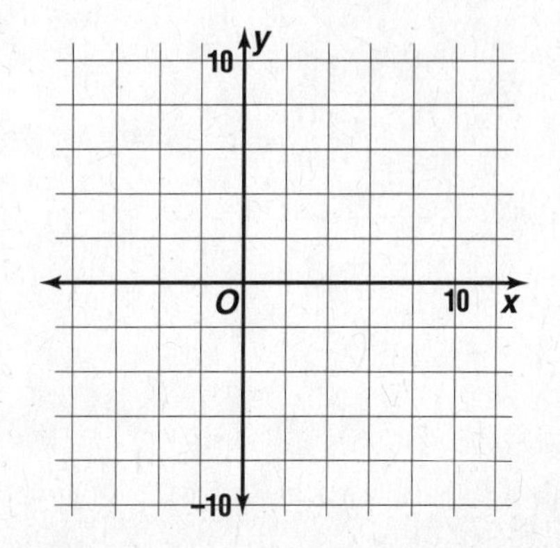